Structural Analysis of Organic Compounds by Spectroscopic Methods

54 TABLES AND 50 EXAMPLES

By

Prof. Dr WILHELM SIMON

and

Dr THOMAS CLERC

Eidgenössische Technische Hochschule
Organic Chemistry Laboratory
Zürich

MACDONALD : LONDON
AMERICAN ELSEVIER : NEW YORK

This English edition is a translation of
a specially revised version of
Strukturaufklärung organischer Verbindungen
mit spektroskopischen Methoden
First published in 1967 by
Akademische Verlagsgesellschaft
Frankfurt am Main

English edition published under licence by
Macdonald & Co. (Publishers) Ltd
49/50 Poland Street, London W.1

Published in United States by
American Elsevier Publishing Company, Inc.
52 Vanderbilt Avenue, New York, New York 10017

Translated by Transcripta Services

British SBN 356 03052 0
American SBN 444 19605 6
Library of Congress Catalog Card Number 74-150948

Printed and bound in Great Britain by
Hazell Watson and Viney Ltd, Aylesbury, Bucks

STRUCTURAL ANALYSIS OF ORGANIC COMPOUNDS BY SPECTROSCOPIC METHODS

Foreword

Instrumental methods, especially infrared, nuclear resonance, mass and electronic spectroscopy, have in recent years become very important for the elucidation of the structure of organic compounds. The number of textbooks and particularly of monographs has increased correspondingly to an alarming degree. The especial effectiveness of the combined use of these four methods for the structural elucidation of organic compounds was first impressively shown in 1963 by R. M. Silverstein and G. C. Bassler in 'Spectrometric Identification of Organic Compounds'. Today this combination is already being taught systematically in various well-known colleges. Experience gained from courses and seminars in the organic chemistry laboratory, Eidgenössiche Technische Hochschule, Zürich, was used in the compilation of the tables given in this book. No claim to comprehensiveness is made; instead data have been selected which by experience have been found sufficient for the interpretation of the spectra of relatively simple compounds. The tables are intended primarily for those who wish to make use of the advances in instrumental analysis without having an expert knowledge of the various spectroscopic methods. Consequently only a basic knowledge, such as today is found in the basic training of an organic chemist, is assumed. There are doubtless omissions in the tables and we should be grateful to have these brought to our attention and also for suggestions for improvements.

We should like to take this opportunity to thank all those who have been in any way involved in the production of this book. We especially thank Dr. C. Pascual, Priv.-Doz. Dr. J. Seibl and J. A. Völlmin for their collaboration in the preparation of tables in their own fields. Likewise we thank the technicians of our instrumental analytical department for recording the spectra.

W. Simon
J. T. Clerc

Zürich, July 1966

Editor's Foreword

The book in this Series *Applications of Spectroscopy to Organic Chemistry* by Brand and Eglinton provides for honours students a review of spectroscopic principles and, by means of examples and problems, shows the relationships between structure and spectra. The successful application of spectroscopic methods depends on experience and practice and this volume by Simon and Clerc is regarded as complementary to the earlier text, in providing a large number of additional examples and problems. It is also likely to be a valuable reference book at research level.

M. F. Grundon

Contents

NMR Tables

MS Tables

Appendix

Introduction

In the following tables abbreviations have been avoided as far as possible. Long explanations are therefore unnecessary and only the following general notes are given.

IR: Infrared spectroscopy

Data given: wave number $\bar{\nu}$ (cm^{-1}) of absorption maxima of solutions in solvents of low polarity such as CCl_4, CS_2, $CHCl_3$ etc. The following abbreviations are used for the assignment of individual absorption bands and for intensity data:

s.	strong
m.	medium
w.	weak
v.	variable
st.	stretching vibration
skel.	skeletal vibration
def.	deformation vibration
as.	asymmetric
sy.	symmetric

UV: Spectroscopy in ultraviolet and visible regions (electronic spectroscopy)

Data given: wavelength λ_{max} (nm) of absorption maxima of ethanolic solutions; decadic logarithm log ε of molar extinction coefficient ε at the corresponding wavelength.

NMR: Proton resonance spectroscopy

Data given: chemical shift δ (ppm) based on tetramethylsilane (TMS) as internal reference in solutions in $CDCl_3$ or CCl_4: Coupling constants J (Hz).

MS: Mass spectroscopy

Data given: mass to charge ratio, *m/e*, of corresponding fragments. The molecular ion is represented by M.

Table 1

Type of compound: **C—C alkane**

	Position	Assignment	Notes
IR	2960...2850 cm^{-1} s.	CH st.	See also: H—(cyclopropyl) : ~3050 cm^{-1} H—C—Hal : ~3010 cm^{-1} CH_3O— : 2830...2815 cm^{-1} CH_3N— : ~2800 cm^{-1} —O—CH_2—O— : ~2780 cm^{-1} H—C(=O)— : 2880...2650 cm^{-1} (usually two bands) Also in same region: =CH— st.; ArH st.; OH assoc. st.
	~1460 cm^{-1} m.	CH_3 def. as. CH_2 def.	'Activated' as with CH_2 groups adjacent to C=C or Ar at 1440...1400 cm^{-1}. → Table 21.
	~1380 cm^{-1} m.	CH_3 def. sy.	Doublet for gem. methyl groups → Table 23. Missing for compounds with no CH_3 groups. In methyl ketones, acetates at 1360...1340 cm^{-1}.
	1350...1100 cm^{-1} v.	C—C st. skel. CH def.	No practical importance except for gem. CH_3 groups → Table 23.
	~720 cm^{-1} v.	**skel.**	Rocking of —CH_2—. Present for —$(CH_2)_{\bar{n}\geq 4}$. For $n<4$ at higher frequencies (up to 800 cm^{-1})—and less intense, often invisible; for cyclohexanes at ~890 cm^{-1}. See also: —O—$(CH_2)_{\bar{n}\geq 4}$: ~740 cm^{-1}. Also in same region: =CH def. (out-of-plane); ArH def. (out-of-plane); NH_2 def. (out-of-plane); NO def.; CS st.; SO st.; PC st.; CCl st.; ring def.; —$OC(CH_3)_3$ skel.

UV-, NMR- and MS- data continued on the next page

Table 1 (cont.)

	Position	Assignment	Notes
UV	none above 210 nm		For saturated hydrocarbons.
NMR	0.8…1.2 ppm 1.1…1.8 ppm	$C-CH_3$ $\left\{\begin{matrix} C-CH_2-C \\ C-CH-C \\ \quad\;\, \vert \\ \quad\;\, C \end{matrix}\right\}$	Characteristic pseudotriplet for $-CH_2-CH_2-CH_3$. Estimation of chemical shift → Table 41. Alicyclics with at least 6-membered rings give usually a broad, unstructured signal. See also: H–(cyclopropyl): **−0.3…0.8 ppm** $\mathbf{J_{CH-CH}}$ **: 6…7 Hz** if freely rotatable; → **Table 44.**
MS		Molecular ion	*n-Alkanes:* weak; *Isoalkanes:* very weak $\Big\}$ $\boldsymbol{\frac{m}{e} = 14\,n + 2}$. *Monocycloalkanes:* moderately intense $\boldsymbol{\frac{m}{e} = 14\,n}$.
		Fragments	*n-Alkanes:* $\frac{m}{e} = 29, 43, 57, 71, 85 \ldots 14n + 1$; continuous variation in intensity; max. at $\frac{m}{e} =$ **43** or **57**; min. at $\frac{m}{e} = M - 15$. *Isoalkanes:* $\frac{m}{e} = 29, 43, 57, 71, 85 \ldots 14n + 1$; irregular variation in intensity; relative max. from fragmentation at branching position. The charge predominantly stays on the more highly branched fragment. Same order of peaks also with aliphatic carbonyl compounds, but these have rearrangement peaks at $\frac{m}{e} = 14n + 2$.

Monocycloalkanes: $\frac{m}{e} = 27, 41, 55, 69, 83 \ldots 14n-1$; relative max. from fragmentation of bond to ring.
Same peak order also with alkenes and alcohols.

Polycycloalkanes: $\frac{m}{e} = 41, 55, 67, 81, 93, 95, 107$;
Same order of peaks also with polyolefines.

Rearrangements

n-Alkanes: none

Isoalkanes: frequently **1→3** rearrangement, $\frac{m}{e} = \mathbf{14n}$:

$$\left[\begin{matrix} H & & C^3 \\ | & & | \\ C^1 & — & C^2 \end{matrix}\right]^{\oplus} \longrightarrow \begin{matrix} H — C^3 \\ \left[C^1 = C^2\right]^{\oplus} \end{matrix}$$

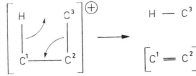

Rearrangement occurs particularly if:
1. C^2: quat. > tert. > sec.
2. C^3: quat. > tert. > sec.
3. C^1: tert. > sec. > prim.

Monocycloalkanes: frequently 1 → 3 rearrangement, $\frac{m}{e} = 14n-2$.

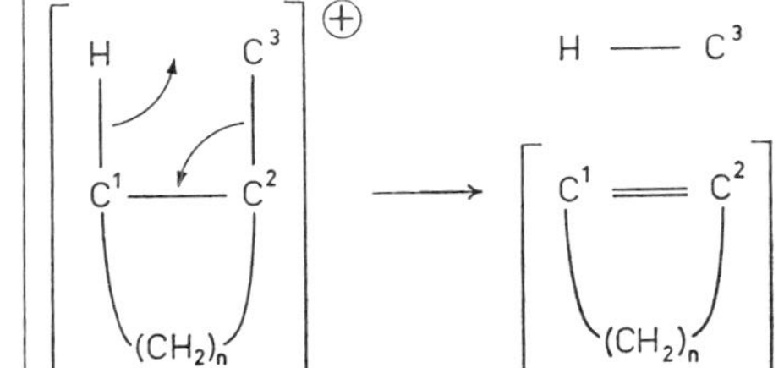

Table 2

Type of compound: **C=C alkene**

	Position	Assignment	Notes
IR	**3100...2975 cm^{-1} m.**	=CH st.	→ Table 24. Often several bands. Missing with tetrasubstituted double bonds. Also in same region: **ArH st.; H—◁ st.; HCHal st.; OH assoz. st.**
	1690...1640 cm^{-1} v.	C=C st.	Less intense if tetrasubstituted, missing at higher symmetry; intense for O—C=C and N—C=C. Often possible to determine type of substitution → Table 24. Also in same region: C=O conjug. st.; NH def. (in-plane); C=N st. NO st. *Enols:* **C=C st.: ~1605 cm^{-1}; C=O st.: ~1640 cm^{-1};** → Table 8. *Conjug. dienes:* 2 bands at ~1650 cm^{-1} and ~1600 cm^{-1}; → Table 29; → Table 30. *Conjug. polyenes:* broad bands at 1650...1580 cm^{-1}; → Table 29; → Table 30.
	1420...1290 cm^{-1} w.	=CH def. (in-plane)	No practical importance.
	990... 675 cm^{-1} s.	=CH def. (out-of-plane)	Sometimes two bands. Missing with tetrasubstituted double bonds. Often possible to determine type of substitution → Table 24. Also in same region; **—$(CH_2)_{n>4}$ skel.; ArH def. (out-of-plane); CH def. (out-of-plane), OH assoz. def. (out-of-plane); C—C skel.; NH_2 def. (out-of-plane); NO def.; CS st.; SO st.; PC st.; CCl st.; Ring def.; —$OC(CH_3)_3$ skel.; C(=O)—H def. (out-of-plane).** See also: CH_2—C=C def.: 1445...1430 cm^{-1}.
UV	none above 210 nm		For isolated double bonds; if highly substituted often end absorption >210 nm. *Conjug. polyenes:* >220 nm ($\log \varepsilon = 4...5$) $\pi \rightarrow \pi^*$; → Table 30. Estimation of λ_{max}: → Table 29.

NMR	4.5...8.0 ppm	C=C–H	Estimation of chemical shift: → Table 42. 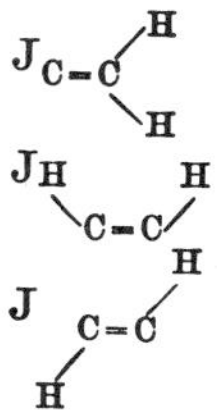 $J_{C=C(H)(H)}$ (geminal) : **0...3,5 Hz** $J_{H-C=C-H}$ (cis) : **5...14 Hz** $J_{H-C=C-H}$ (trans) : **12...18 Hz** $J_{CH-C=CH}$: **0...3 Hz** The coupling constants are dependent on electronegativity of the substituents at the double bond; → Table 44. See also: CH_3—C=C: 1.6...1.8 ppm; CH_2—C=C: 1.7...2.4 ppm.
MS		Molecular ion	Moderately intense $\frac{m}{e} = 14n$.
		Fragments	$\frac{m}{e} = \mathbf{27, 41, 55, 69, 83, \ldots 14n - 1}$; in many cases base peak formed by fragmentation between α- and β-position to double bond. Care: double bond readily migrates! Same order of peaks for monocycloalkanes and alcohols.
		Rearrange- ments	Frequently 1 → 3 rearrangement, $\frac{m}{e} = 14n - 2$. 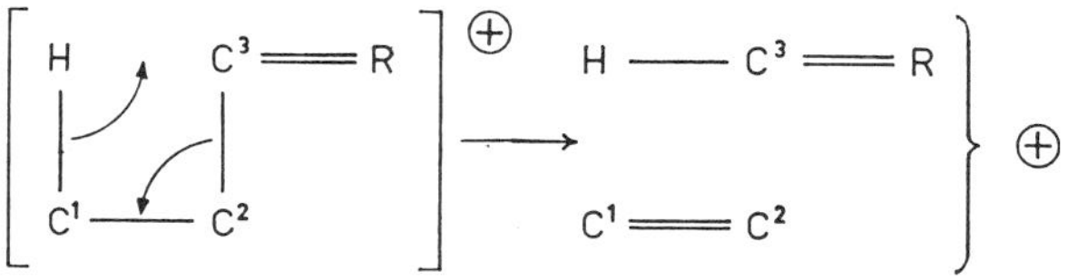 $[H, C^3{=}R, C^1{-}C^2]^{\oplus} \rightarrow \{H{-}C^3{=}R,\ C^1{=}C^2\}^{\oplus}$ Formation of preferentially stabilised ions. In cyclohexenes often also retro-Diels-Alder reaction: 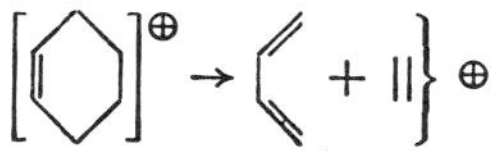 [cyclohexene]$^{\oplus}$ → {butadiene + ethylene}$^{\oplus}$ Formation of preferentially stabilised ions.

Table 3

Type of compound: **C≡C alkyne**

	Position	**Assignment**	Notes
IR	~3300 cm^{-1} w.	≡CH st.	Very sharp; missing with disubstituted triple bonds. Also in same region: OH assoc. st.; NH st.
	2260...2100 cm^{-1} w.	C≡C st.	Usually sharp; missing at higher symmetry. Also in same region: X≡Y st.; X=Y=Z st.; → Table 22.
	700...600 cm^{-1}	≡CH def. (out-of-plane)	Missing with disubstituted triple bonds; no practical value.
			See also: CH_2—C≡C def.: 1445...1430 cm^{-1}.
UV	>210 nm end absorption		Often several weak bands < **240** nm (log ε = 1...2)
NMR	2.0...3.2 ppm	C≡C—H	$J_{CH—C≡CH}$: 2...3 Hz See also: CH_3—C≡C: ~1.8 ppm CH_2—C≡C: ~2.0 ppm
MS		Molecular ion	Weak; usually missing with 1-alkynes.
		Fragments Rearrangements	Fluctuating between aromatic and alkane types.

Table 4

Type of compound: **aromatic**

	Position	Assignment	Notes
IR	3080...3030 cm^{-1} v.	ArH st.	Also in same region: =CH st.; H–(cyclopropyl) st.; HCHal st.; OH assoc. st.
	2000...1660 cm^{-1} w.	Combinations and overtones	So-called benzene finger. Only clear at higher concentration and/or greater cell thicknesses. Substitution type often determinable; → Table 25.
	1625...1575 cm^{-1} v. 1525...1475 cm^{-1} v.	skel. skel.	Bands often split. Bands at ~1500 cm^{-1} sometimes missing, but usually more intense than bands at ~1600 cm^{-1}. Also in same region: C=C conjug. st.
	1460...1440 cm^{-1} v.	skel.	
	1225...950 cm^{-1} w.	ArH def. (in-plane)	Often several bands; intense in presence of polar substituents, no practical importance.
	900...735 cm^{-1} m. s.	ArH def. (out-of-plane), skel.	1 to 3 bands. Substitution type often determinable; → Table 25. Also in same region; $-(CH_2)_{\overline{n}\geqslant 4}$ skel.; =CH def. (out-of-plane); C–O–C (epoxide) skel.: NH_2 def. (out-of-plane); NO def.; CS st.; SO st.; PC st.; CCl st.; Ring def.; $-OC(CH_3)_3$ skel.; H–C(=O) def. (out-of-plane). See also: CH_2–Ar def.: 1445...1430 cm^{-1}.

UV-, NMR- and MS- data continued on the next page

Table 4 (cont.)

	Position	Assignment	Notes
UV	**205…260 nm** (log $\varepsilon = \sim 4$) **260…300 nm** (log ε = 2.5…3.5)		→ Table 34, → Table 35. → Table 36. Conjugation raises ε and shifts λ_{max} to higher wavelengths.
NMR	6.5…8.5 ppm	**ArH**	Estimation of chemical shift: → Table 42. **J_{ortho} : 7…10 Hz** **J_{meta} : 2…3 Hz** **J_{para} : 0…1 Hz** See also: CH_3—Ar: 2.2…2.5 ppm singlet C—CH_2—Ar: 2.3…2.8 ppm
MS		Molecular ion	Very intense, often base peak.
		Fragments	$\frac{m}{e}$ = 39, 50…53, 63…65, 75…78; higher substitution causes shift to lower values. Doubly charged ions frequent. Depending on substitution the following intense peaks appear: C_6H_5—CH : $\frac{m}{e}$ = **90…92** C_6H_5—N : $\frac{m}{e}$ = **91…93** C_6H_5—O : $\frac{m}{e}$ = **93…94** C_6H_5—C=O : $\frac{m}{e}$ = **105** naphthyl : $\frac{m}{e}$ = **127** biphenyl : $\frac{m}{e}$ = **153**

	Rearrangements	McLafferty rearrangement occurs with certain *o*-substituted aromatic compounds: 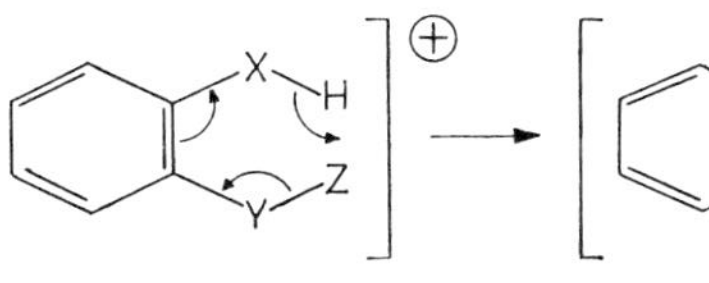 e.g. X: CH_2, O, NR etc. YZ: C(=O)—OR, C—Hal etc. Aromatic compounds containing N: HCN splits off.

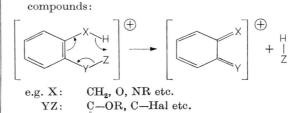

Table 5

Type of compound: **C—O—C ether**

	Position	Assignment	Notes
IR	1275...1020 cm^{-1} s.	C—O st.	1150...1070 cm^{-1}: $-\overset{\mid}{\underset{\mid}{C}}-O-\overset{\mid}{\underset{\mid}{C}}-$ st. as. 1275...1200 cm^{-1}, 1075...1020 cm^{-1} } =C—O—C; Ar—O—C {st. as., st. sy. Also in same region: C—O st. of alcohols, carboxylic acids, esters NO st.; C=S st.; CF st. See also: $-O(CH_2)_{n>4}$: ~ 740 cm^{-1} w. $-OC(CH_3)_3$: 920...800 cm^{-1} s. $-OCH_3$: 2830...2815 cm^{-1} $-O-CH_2-O-$: ~ 2780 cm^{-1}
UV	none above 210 nm		
NMR	3.3...4.0 ppm	CH_3-O $-CH_2-O-$	Singlet See also: CH_3OAr: ~3.8 ppm, singlet $C-CH_2OAr$: 4.1...4.6 ppm H_2C(O)(O)-benzo (methylenedioxybenzene) : ~ 6.0 ppm, very sharp singlet.
MS		Molecular ion Fragments	*Aliphatic:* small; M + 1 often more intensive than M. *Aromatic:* intense. $\frac{m}{e}$ = 31, 45, 59, 73...14n + 3; base peak usually formed in aliphatic ethers by fragmentation in β-position to oxygen: $R_1 \vert C-O-R_2$.

	Rearrange-ments	Aliphatic ethers frequently give 1 → **3** rearrangement: 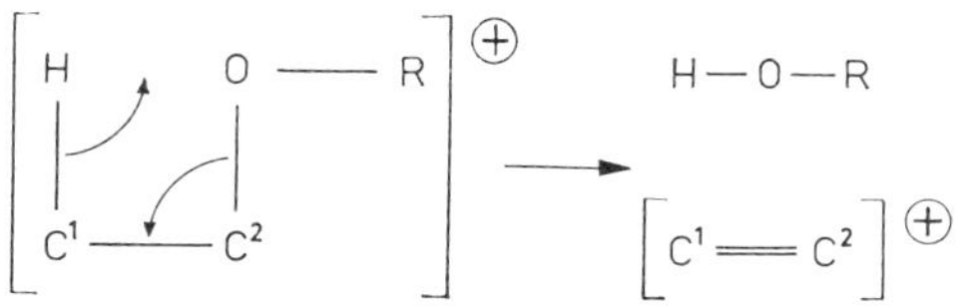 Aromatic ethyl and higher ethers give McLafferty rearrangement: 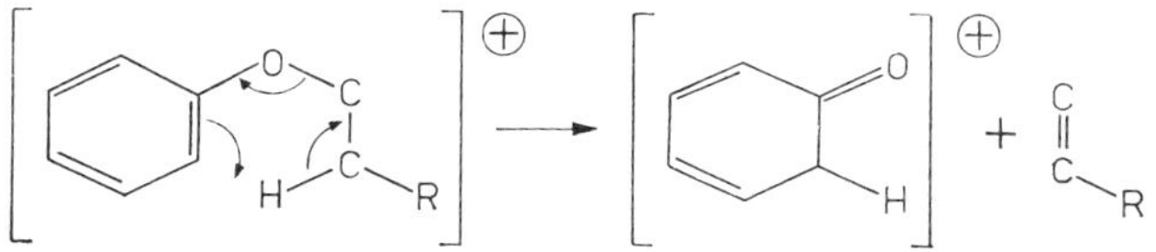 Phenolic ethers: CO splits off.

Table 6

Type of compound: **C—OH alcohol, carboxylic acid**

	Position	Assignment	Notes
IR	3670...2500 cm^{-1} v.	OH st.	3670...3500 cm^{-1}: *free OH*. Sharp band. In so-called 'nonpolar' solvents (CCl_4, $CHCl_3$ etc.): ~3640 cm^{-1}: prim. alcohols. ~3630 cm^{-1}: sec. alcohols. ~3620 cm^{-1}: tert. alcohols. ~3610 cm^{-1}: phenols; → Table **4**. ~3550 cm^{-1}: carboxylic acids; → Table 10. 3600...3200 cm^{-1}: *intermolecular assoc. OH*. Relative intensity dependent on concentration. 3600...3450 cm^{-1}: Dimer. Very sharp band. H-bonds with solvent also absorb in this region. 3400...3200 cm^{-1}: Polymers. Broad bands. In the spectra of liquids and solids usually only these bands appear. 3600...2500 cm^{-1}: *intramolecular assoc. OH*, *chelates*. Relative intensity independent of concentration. 3600...3500 cm^{-1}: O—H...O—H etc. 3200...2500 cm^{-1}: O—H...O=C etc. broad band, often difficult to detect. Also in same region: NH st.; CH st.; overtone of C=O st.; H_2O.
	1500...1250 cm^{-1} m.	OH def. (in-plane)	No practical value.
	1300...1000 cm^{-1} s.	C—O st.	Very intense. Position very sensitive to branching, substitution and double bond in α-position. ~1250 cm^{-1}: carboxylic acids, → Table 10. ~1200 cm^{-1}: phenols, → Table 4. ~1150 cm^{-1}: tert. alcohols. ~1100 cm^{-1}: sec. alcohols. ~1050 cm^{-1}: prim. alcohols.

			Also in same region: C—O st. of ethers and esters, NO st.; C=S st.; CF st.
UV	none above 210 nm		
NMR	0.5…16 ppm	—OH	→ Table 47. *Alcohols* usually: 0.5…5.5 ppm. *Phenols* usually: 4.5…7.0 ppm; see H—Ar: 6.5…8.5 ppm. *Carboxylic acids*: 9.5…13 ppm; see HC—COO: 2.0…2.5 ppm. *Chelates*, *Enols*: 10…16 ppm.
MS		Molecular ion	*Alcohols:* small, often missing for highly branched and primary alcohols. In this case the peaks with the highest mass usually are at M-18 and/or M-15. *Phenols:* intense; sometimes M-1 is more intense than M.
		Fragments	*Alcohols:* $\frac{m}{e}$ = **31, 45, 59…; M-46, M-33, M-18.** Frequently: prim. alcohols: $\frac{m}{e} = \mathbf{31} > \frac{m}{e} = \mathbf{45}, \quad \frac{m}{e} = \mathbf{59}$ sec. alcohols: $\frac{m}{e} = \mathbf{31} < \frac{m}{e} = \mathbf{45} > \frac{m}{e} = \mathbf{59}$ tert. alcohols: $\frac{m}{e} = \mathbf{31}, \quad \frac{m}{e} = \mathbf{45} \ll \frac{m}{e} = \mathbf{59}.$ Similar order of peaks as with alkenes and monocycloalkanes. *Phenols:* $[C_6H_5\text{—}O]^{\oplus}$; frequently metastable peaks caused by splitting of 28 (CO) and 29 (HCO).
		Rearrangements	*Alcohols:* water splits off M-18. *Phenols:* CO splits off. McLafferty rearrangement occurs if suitable substituent in the *o*-position: $[\text{o-substituted phenol (O—H, Y—Z)}]^{\oplus} \longrightarrow [\text{cyclohexadienone (=O, =Y)}]^{\oplus} + \text{H—Z}$ e.g. Y—Z: **C(=O)—OR; C—Hal; OR** etc.

Table 7

Type of compound: **C—C(=O)—H aldehyde**

	Position	Assignment	Notes
IR	2880…2650 cm^{-1} m.	C(=O)—H st.	Often 2 bands at ~2820 cm^{-1} and ~2720 cm^{-1}. Also in same region: CH_3O st.; CH_3N st.; —O—CH_2—O— st.
	1730…1650 cm^{-1} s.	C=O st.	~1730 cm^{-1}: aliphatic aldehydes. ~1700 cm^{-1}: aromatic aldehydes; → Table 4. ~1690 cm^{-1}: $\alpha\beta$-unsat. aldehydes } → Table 2 ~1675 cm^{-1}: $\alpha\beta$, $\gamma\delta$-unsat. aldehydes } → Table 2 ~1660 cm^{-1}: H—C=O·····H; → Table 6 or Table 12. Also in same region: C=O st. of other carbonyl compounds; C = C st.; C=N st.; NO st.
	1440…1160 cm^{-1} m.		Often several bands; no practical value.
	975… 780 cm^{-1} w.	CHO def. (out-of-plane)	No practical value. See also: CH_2—C=O: 1440…1400 cm^{-1}.
UV	270…300 nm (log ε = 1…1.5)	$n \to \pi^*$	Often difficult to detect because of low intensity. *$\alpha\beta$-unsat. aldehydes:* → Table 31. 210…235 nm (log ε = 3.7…4.3); $\pi \to \pi^*$. 310…325 nm (log ε = 1…1.8); $n \to \pi^*$. *Mononuclear aromatic aldehydes:* 250 nm (log ε = 4…4.5); → Table 36. 280 nm (log ε = ~3); usually hidden by band at ~250 nm.
NMR	2.0…2.5 ppm 9.4…10 ppm	CH_2—C=O C(=O)—H	J_{CH-CHO}: 1…3 Hz.

			αβ-unsat. aldehydes: H—C=C—C=O: 6.5...7.2 ppm. aromatic aldehydes: 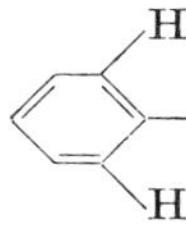 (ortho-H)—C=O: ~8.0 ppm.
MS		Molecular ion	Present; moderately intense for aliphatic aldehydes; intense for aromatic aldehydes, accompanied by an equally intense peak at M-1.
		Fragments	$\frac{m}{e}$ = M-1, M-18, M-28,...; for aliphatic aldehydes the base peak often arises from McLafferty rearrangement, for aromatic aldehydes usually M or M-1.
		Rearrangements	*Aliphatic aldehydes:* McLafferty rearrangement: 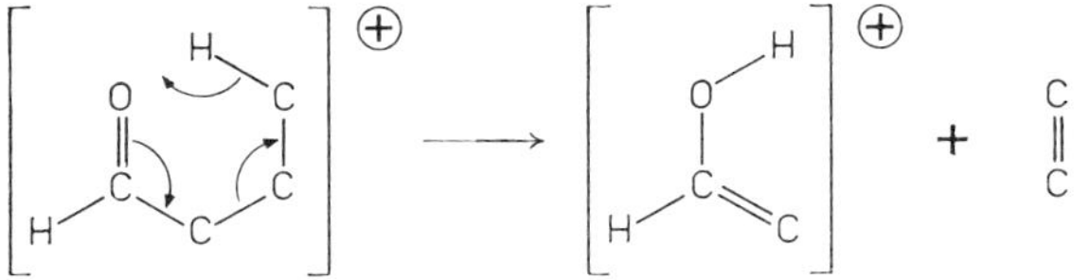 $[H{-}C(=O){-}C{-}C{-}C{-}H]^{\oplus} \longrightarrow [H{-}C(OH){=}C]^{\oplus} + C{=}C$ If unsubstituted in the α-position: $\frac{m}{e}$ = **44**; otherwise correspondingly higher. *Aromatic aldehydes:* McLafferty rearrangement if suitable *o*-substitution. 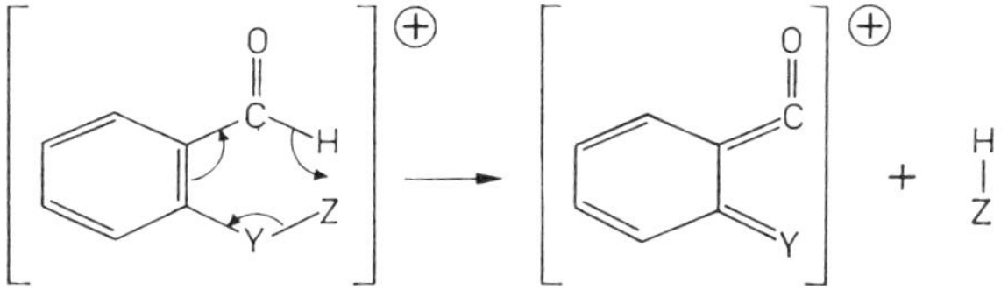 $[C_6H_4(CHO)(Y{-}Z{-}H)]^{\oplus} \longrightarrow [C_6H_4(=C=O)(=Y)]^{\oplus} + H{-}Z$ e.g. YZ: C(=O)—OR; C—Hal; O—R etc.

Table 8

Type of compound: **C—C(=O)—C ketone**

	Position	Assignment	Notes
IR	1780...1550 cm^{-1} s.	C=O st.	*1780...1700 cm^{-1}:* —C—C(=O)—C— ~ 1780 cm^{-1}: 4-membered-ring ketones. ~ 1755 cm^{-1}: α, α'-dihalogenoketones. ~ 1745 cm^{-1}: 5-membered-ring ketones. ~ 1735 cm^{-1}: α-halogeno-; α, α-dihalogenoketones. ~ 1720 cm^{-1}: α-; β-diketones; sometimes doublet. ~ 1715 cm^{-1}: aliph. ketones; γ-diketones; 6-membered ring ketones. ~ 1705 cm^{-1}: ring ketones with $n \geqslant 7$. *1700...1645 cm^{-1}:* O=C—C=C. ~ 1690 cm^{-1}: aryl ketones; → Table 4. ~ 1675 cm^{-1}: α, β-unsat. ketones; → Table 2, 1,2-; 1,4-quinones. ~ 1665 cm^{-1}: α β, $\alpha'\beta'$-; $\alpha\beta$, $\gamma\delta$-unsat. ketones; → Table 2, diaryl ketones; → Table 4. *1675...1550 cm^{-1}:* C=O····H; → Table 6 or Table 12. ~ 1675 cm^{-1}: enolised α-diketones; C=C st.: ~1650 cm^{-1} s. ~ 1650 cm^{-1}: enolised β-diketones. ~ 1615 cm^{-1}: chelated β-diketones; C=C st.: ~1605 cm^{-1} s. In same region also: C=O st. of other carbonyl compounds; C=C st.; C=N st.; NO st.
	1325...1075 cm^{-1} m.		Often several bands; no practical value.
UV	270...300 nm (log ε = 1...1.5)	$n \to \pi^*$	Often difficult to detect because of low intensity. *α-diketones:* 270...300 nm (log ε = 1...1.5) 330...470 nm (log ε = 1...1.5) *$\alpha\beta$-unsat. ketones:* → Table 31. 210...250 nm (log ε = 3.7...4.3) $\pi \to \pi^*$. 310...325 nm (log ε = 1...1.8) $n \to \pi^*$.

			Estimation of λ_{max} for $\alpha\beta$-unsat. ketones → Table 33. *Mononuclear aromatic ketones:* 250 nm ($\log \varepsilon = 4...4.5$); → Table 36. 280 nm ($\log \varepsilon = \sim 3$) usually hidden by the band at 250 nm.
NMR	2.0...2.5 ppm	$CH_2{-}C{=}O$ $CH_3{-}C{=}O$	Singlet. $\alpha\beta$-unsat. ketones: H—C=C—C=O: 6.5...7.2 ppm. aromatic ketones: 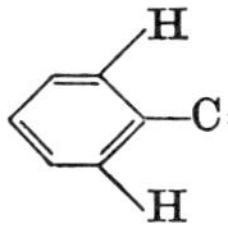 (ring H ortho to C=O): ~8 ppm.
MS		Molecular ion **Fragments**	Present; very intense for aromatic ketones. Preferential fragmentation at bond adjacent to carbonyl group: $R_1{-}CO{-}R_2$. For aromatic ketones this usually gives a base peak: $R_1{-}CO{-}C_6H_5$ For unsubstituted fragments: $\frac{m}{e} = 105$; otherwise correspondingly higher.
		Rearrangements	Aliphatic ketones: McLafferty rearrangement. If α-position unsubstituted: 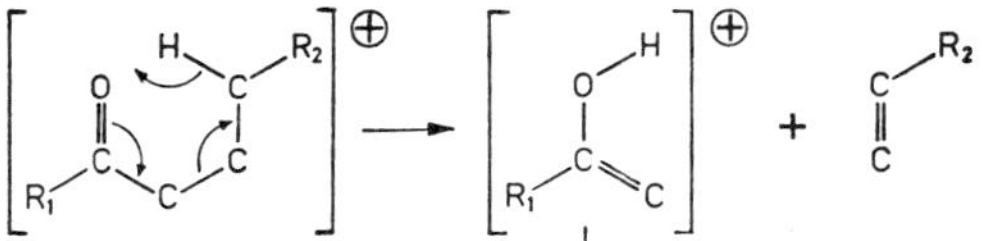 methyl ketones: $\frac{m}{e} = 58$. ethyl ketones: $\frac{m}{e} = 72$, otherwise correspondingly higher. 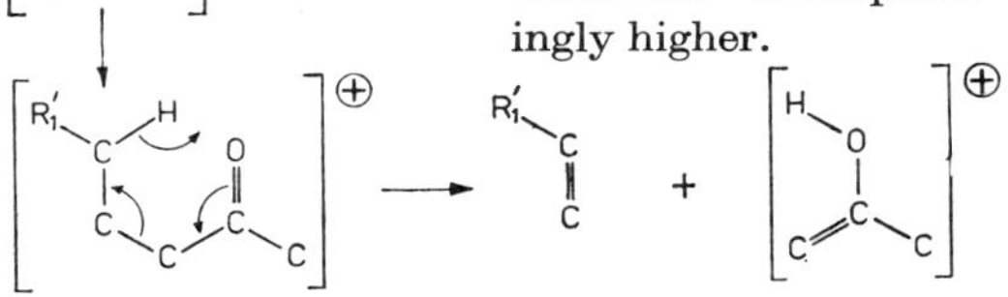

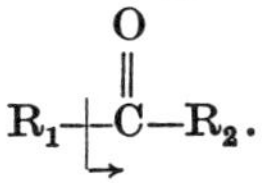

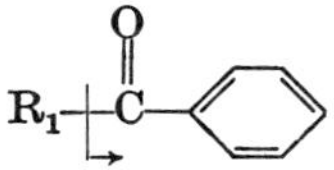

Table 9

Type of compound: **C—C(=O)—O—** **ester, lactone**

	Position	Assignment	Notes
IR	1880...1635 cm^{-1} s.	C=O st.	~1880 cm^{-1}: $\beta\gamma$-unsat. γ-lactones; → Table 2. ~1825 cm^{-1}: β-lactones. ~1770 cm^{-1}: vinyl esters; → Table 2. phenyl esters; → Table 4. γ-lactones ~1760 cm^{-1}: $\gamma\delta$-unsat. δ-lactones; → Table 2. ~1750 cm^{-1}: $\alpha\beta$-unsat. γ-lactones; → Table 2. If α—H present: 2 bands at ~1755 cm^{-1} and ~1785 cm^{-1}. ~1745 cm^{-1}: α-ketoesters ArCOOAr; → Table 4. ~1735 cm^{-1}; aliph. esters δ-lactones β-ketoesters. ~1720 cm^{-1}: $\alpha\beta$-unsat. esters; → Table 2. aromat. esters; → Table 4. $\alpha\beta$-unsat. δ-lactones; → Table 2. 1670...1635 cm^{-1}: OC=O·····H; → Table 6 or Table 12. Also in same region: C=O st. of other carbonyl compounds; C=C st.; C=N st.; NO st.
	1300...1050 cm^{-1} s.	C—O st.	2 intense bands: C—O st. as. and C—O st. sy.; st. as. at higher wave number and more intense than st. sy.

			Approx. position for st. as.: $\sim$1260 cm^{-1} $\alpha\beta$-unsat. esters; → Table 2. aromatic esters; → Table 4. $\sim$1240 cm^{-1}: acetates; see: CH_3COO: 1380...1365 cm^{-1}; NMR: CH_3COO: $\sim$2.1 ppm singlet. $\sim$1210 cm^{-1}: vinyl esters; → Table 2. phenyl esters; → Table 4. $\sim$1185 cm^{-1}: formates, propionates, higher aliphatic esters. $\sim$1180 cm^{-1}: γ-, δ-lactones $\sim$1165 cm^{-1}: methyl esters NMR: CH_3OCO: $\sim$3.7 ppm singlet. Also in this region C—O st. of alcohols, ethers, carboxylic acids; NO st.; C=S st.; C—F st. See also: $-O-(CH_2)_{\overline{n} \geqslant 4}$: $\sim$ 740 cm^{-1} $-OC(CH_3)_3$: 920... 800 cm^{-1} s. $CH_2C=O$: 1440...1400 cm^{-1}
UV	none above 210 nm		Esters of $\alpha\beta$-unsat. acids: < 230 nm ($\log \varepsilon = \sim 4$): $\pi \rightarrow \pi^*$; → Table 32.
NMR	2.1...2.6 ppm 4.0...4.5 ppm	CH_2-COO CH_2-OOC	See also: CH_3COO: 2.0...2.6 ppm singlet. CH_3OOC: 3.5...4.0 ppm singlet.

MS data continued on the next page

Table 9 (cont.)

	Position	Assignment	Notes
MS		Molecular ion	Present; intense for aromatic esters; M + 1 sometimes more intense than M.
		Fragments	Preferential fragmentation at bond next to carbonyl group: $R_1-\vert-C(=O)-O-R_2$; $R_1-C(=O)-\vert-O-R_2$.
		Rearrangements	Aliphatic esters: McLafferty rearrangement: $[RO-C(=O)-C-C-C-H]^{\oplus} \longrightarrow [RO-C(-O-H)=C]^{\oplus} + C{=}C$ If acids unsubstituted in α-position methyl esters $\frac{m}{e} = 74$ ethyl esters $\frac{m}{e} = 88$ otherwise correspondingly higher. $[R-C(=O)-O-C-C-H]^{\oplus} \longrightarrow [R-C(-O-H)=O]^{\oplus} + C{=}C$ For acetates $\frac{m}{e} = 60$

Fragments	*Aliphatic:* $\frac{m}{e} = 31, 44, 45;\ M - 45, M - 18, M - 17.$ *Aromatic:* $\frac{m}{e} = M - 45, M - 44, M - 17.$
Rearrangement	*Aliphatic:* McLafferty rearrangement: 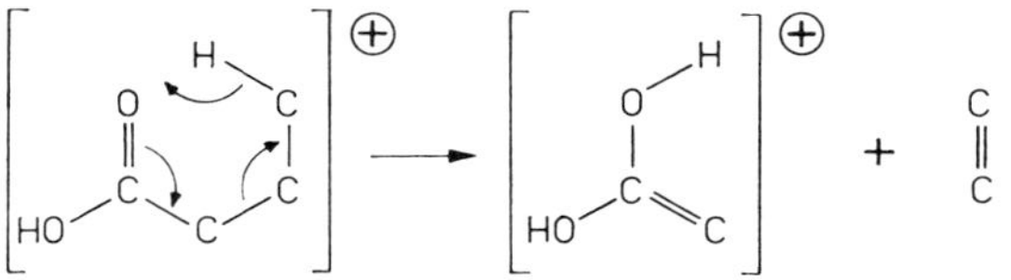 If α-position unsubstituted: $\frac{m}{e} = 60$, otherwise correspondingly higher. *Aromatic:* if *o*-substituent containing H present, $\frac{m}{e} = M - 18$ instead of $\frac{m}{e} = M - 17$: 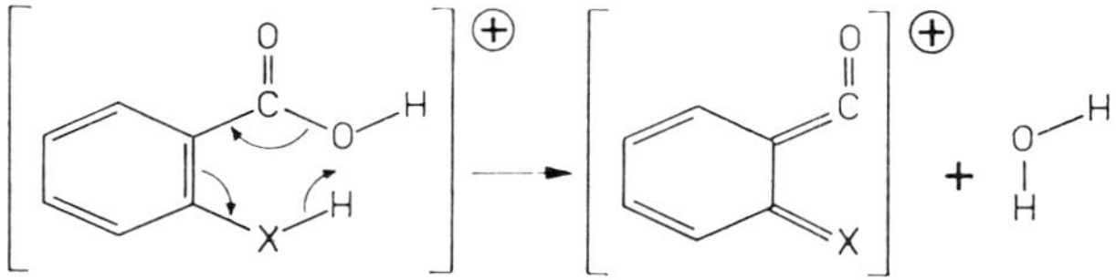 $\frac{m}{e} = M - 18$

Table 11

Type of compound: **C—C—N amide, lactam** (with **O** double-bonded (‖) to the middle C)

	Position	Assignment	Notes
IR	3500...3000 cm^{-1} v.	NH st.	3500...3400 cm^{-1}: *Free.* prim. amides: 2 bands. sec. amides: 1 band at ~3400 cm^{-1}. tert. amides: missing. 3300...3000 cm^{-1}: *Associated.* prim. amides: several bands at 3200...3000 cm^{-1}. sec. amides: 2 bands at ~3300 cm^{-1} and ~3070 cm^{-1}. lactams: 2 bands at ~3175 cm^{-1} and ~3070 cm^{-1}. tert. amides: missing. Also in same region: =CH st.; OH; NH st. of other NH-compounds overtone of C=O st.; ≡CH st.; H_2O.
	1690...1650 cm^{-1} s.	'Amide I'	prim. amides: *free*: ~1690 cm^{-1}; *assoc.*: ~1650 cm^{-1}. sec. amides: *free*: ~1680 cm^{-1}; *assoc.*: ~1655 cm^{-1}. tert. amides: *free*: ~1650 cm^{-1}; *assoc.*: ~1650 cm^{-1}. Also in same region: C=O st. of other carbonyl compounds; C=C st.; C=N st.; NO st.; C=S st.
	1640...1530 cm^{-1} s.	'Amide II'	prim. amides: *free*: ~1600 cm^{-1}; *assoc.*: ~1640 cm^{-1}. sec. amides: *free*: ~1530 cm^{-1}; *assoc.*: ~1550 cm^{-1}. tert. amides: missing. Also in same region: C=C st.; Ar skel.; NH def. of other NH-compounds; C=O·····H st.; NO st. See also: prim. amides: ~1410 cm^{-1}. sec. amides: 1300...1200 cm^{-1} 'Amide III' 720... 620 cm^{-1} 'Amide IV' CH_2—C=O: 1440...1400 cm^{-1} CH_3—N: 2820...2760 cm^{-1}

UV	none above 210 nm		Amides of $\alpha\beta$-unsat. acids: <230 nm ($\log \varepsilon = 3 \ldots 4$) $\pi \rightarrow \pi^*$
NMR	2.0…2.6 ppm	$CH_2{-}C({=}O){-}N$	See also: $CH_3{-}C({=}O){-}N$: ~1.9 ppm singlet
	3.2…3.8 ppm	$C({=}O){-}N{-}CH_2$	$C({=}O){-}N{-}CH_3$: ~2.9 ppm singlet
	5.0…8.5 ppm	$C({=}O){-}NH$	→ Table 47. Signals usually very broad, often only recognisable from integral.
MS		Molecular ion	Present; intense for aromatic amides. If number of nitrogen atoms is odd, then M is odd.
		Fragments	Preferred fragmentation: 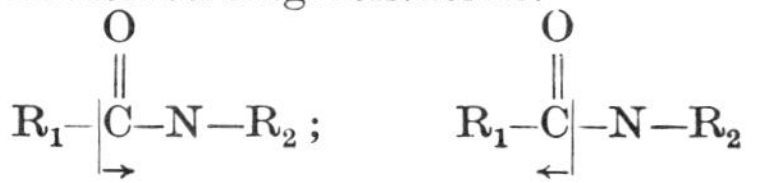 $R_1{-}\vert C({=}O){-}N{-}R_2$; $R_1{-}C({=}O)\vert{-}N{-}R_2$
		Rearrangements	1 → 3 and McLafferty rearrangements frequent: 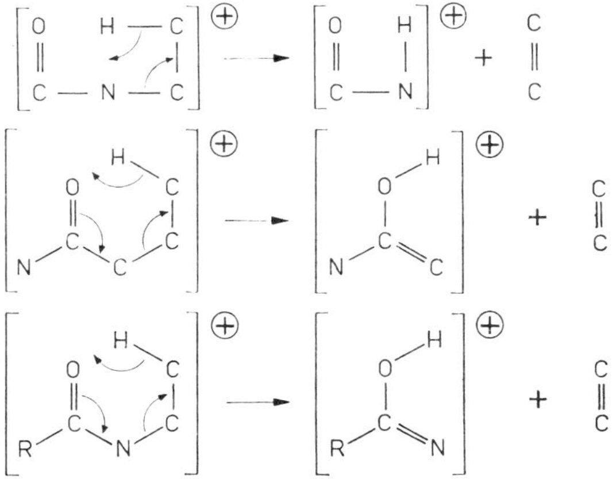

$$[O{=}C{-}N(H{-}C){-}C]^{\oplus} \longrightarrow [O{=}C{-}N{-}H]^{\oplus} + C{=}C$$

$$[N{-}C({=}O{\cdots}H){-}C{-}C{-}C]^{\oplus} \longrightarrow [N{-}C(O{-}H){=}C]^{\oplus} + C{=}C$$

$$[R{-}C({=}O{\cdots}H){-}N{-}C{-}C]^{\oplus} \longrightarrow [R{-}C(O{-}H){=}N]^{\oplus} + C{=}C$$

Rest of MS data continued on the next page

Table 11 (cont.)

	Position	Assignment	Notes
			For *o*-substituted aromatic amides: [C₆H₄(C(=O)–N)(X–H)]⁺ → [C₆H₄(=C=O)(=X)]⁺ + N–H e.g. X: CH_2, O, NR etc.

Table 12

Type of compound: **C—N amine**

	Position	Assignment	Notes
IR	3500...3100 cm^{-1} v.	NH st.	prim. amines: usually 2 bands at ~3500 cm^{-1} and ~3400 cm^{-1}; overtone of NH def. at 3200 cm^{-1}. sec. amines: usually one band; less intense in N-heterocycles. tert. amines: missing. Also in same region: NH st. of other NH-compounds; OH st.; =CH st. overtone of C=O st.; H_2O.
	1640...1490 cm^{-1} v.	NH def. (in-plane)	prim. amines: sec. amines: tert. amines: missing. Also in same region: C=C st.; Ar skel.; NH def. of other NH-compounds; NO st.
	1360...1030 cm^{-1} m.	C—N st.	No practical value; tert. and aromatic amines give two bands.

	900… 650 cm^{-1} w.	NH def. (out-of-plane)	Broad; no practical value. Missing in tert. amines. See also: CH_3—N(aliph.)$_2$ } CH st. one band at ~2800 cm^{-1}. $(CH_3)_2N$—Ar. } $(CH_3)_2N$—aliph. CH st. 2 bands at 2820 cm^{-1} and 2770 cm^{-1}
UV	< 220 nm (log ε = ~ 2.5)	$n \rightarrow \sigma^*$	No practical value.
NMR	2.2…2.7 ppm 0.5…6.0 ppm	CH_2—N C—NH	See also: CH_3—N: ~2.3 ppm singlet → Table 47.
MS		Molecular ion	Usually missing for aliphatic amines; intense for aromatic amines. If number of nitrogen atoms is odd, then M is odd.
		Fragments	Aliphatic amines: $\frac{m}{e} = 30$; many even-numbered ions. Preferred fragmentation: N–C|–R Aromatic amines: $[C_6H_5\text{—N}]^{\oplus}$
		Rearrangements	Aliphatic amines: 1 → 3 rearrangement frequent: $[H—C / N—C]^{\oplus} \longrightarrow [H—N]^{\oplus} + C{=}C$

Table 13

Type of compound: $-\underset{|}{\overset{NH_2}{C}}-\overset{O}{\overset{\|}{C}}-OH$ **α-amino acid (zwitterion, hydrochloride)**

	Position	Assignment	Notes
IR	3130...2500 cm^{-1} m. 2150...2000 cm^{-1} w.	NH_3^+ st.	Several bands or a broad band. *Zwitterion:* often missing.
	1755...1560 cm^{-1} s.	C=O st.	*Zwitterion:* $\leqslant 1600\ cm^{-1}$ *Hydrochloride:* $\geqslant 1730\ cm^{-1}$
	1660...1590 cm^{-1} w.	NH_3^+ def. as.	*Zwitterion:* $\geqslant 1610\ cm^{-1}$ *Hydrochloride:* $\leqslant 1610\ cm^{-1}$
	1550...1485 cm^{-1} v. 1335...1300 cm^{-1} m.	NH_3^+ def. sy.	
UV	none above 210 nm		
NMR	~3.8 ppm	H–C(–COO)–N<	in D_2O

MS

Free α-amino acids are unsuitable for mass spectrometric investigations because of their thermal instability and low volatility. The following data are for ethyl esters of α-amino acids.

Molecular ion	very weak, nearly always accompanied by M + 1; intensity relation is very dependent on conditions.
Fragments	$\frac{m}{e} = 30, 44\ldots$ Many even-numbered ions. Most important fragmentations: R–|–CH(NH₂)–COOC₂H₅ (→) $\frac{m}{e} = 102$ R–CH(NH₂)–|–COOC₂H₅ (←) $\frac{m}{e} = M - 73$

	Rearrangements	$1 \rightarrow 3$ and McLafferty rearrangements frequently follow simple fragmentation:

$$\left[\begin{array}{c} COOC_2H_5 \\ | \\ C-NH_2 \\ | \\ R \end{array}\right]^{\oplus} \longrightarrow \left[\begin{array}{c} C=NH_2 \\ | \\ R \end{array}\right]^{\oplus}$$

$$\left[R'-C(H)-C-\overset{H}{C}=NH_2\right]^{\oplus} \longrightarrow \left[H-CH=NH_2\right]^{\oplus} \quad \frac{m}{e}=30 \qquad R'-C=C$$

$$\left[C(H)-C-CH_2-CH=NH_2\right]^{\oplus} \longrightarrow C=C \; + \; \left[CH_2=CH-NH_3\right]^{\oplus} \quad \frac{m}{e}=44$$

Table 14

Type of compound: **C, H, S compounds**

	Position	**Assignment**	Notes
IR	2600...2550 cm^{-1} w.	S—H st.	At lower wave numbers if associated.
	1200...1050 cm^{-1} s.	C=S st.	C(=S)—N: 1550...1460 cm^{-1} corresponding to 'Amide II', missing if N tertiary. 1300...1100 cm^{-1} corresponding to 'Amide I'.
	800... 570 cm^{-1} w.	C—S st.	
	500... 450 cm^{-1} w.	S—S st.	See also: CH_2—S: 2700...2630 cm^{-1}, ~1420 cm^{-1}; CH_3—S: 1325...1300 cm^{-1}
UV	⩽ 250 nm (log ε = ~2.5)	$n \rightarrow \sigma^*$	C(=S)—N: ~250 nm (log ε = ~4)
NMR	1...2 ppm	SH	→ Table 47. Thioenols: 3...5 ppm, usually broad. $J_{CH\text{-}SH}$: ~8Hz. See also: CH_3S: 2.0...2.4 ppm singlet; CH_2S: 2.4...3.0 ppm
MS		Molecular ion	Present. Characteristic: isotope peak at M + 2: 4.4* number of S atoms gives approx. intensity as percentage of peak at M.
		Fragments	$\frac{m}{e}$ = 33, 34, 47, 48, 61, 75...; M − 34. Fragmentation essentially similar to that of corresponding oxygen compounds.
		Rearrangements	Essentially similar to the corresponding oxygen compounds.

Table 15

Type of compound: **S, O compounds**

	Position	Assignment	Notes
IR	1400...1310 cm^{-1} s.	SO_2 st. as.	Tosylate: 1360 cm^{-1} broad.
	1230...1040 cm^{-1} s.	SO_2 st. sy.	1170 cm^{-1} sharp.
		S=O st.	
	900... 700 cm^{-1} s.	S—O st.	
UV	$\leqslant$ 230 nm		SO_2: <210 nm SO : 210...230 nm (log ε = ~3.6)
NMR	~2.8 ppm	CH_2—SO	
	2.5...3.5 ppm	CH_2—SO_2	CH_3SO_2: ~2.5 ppm singlet
MS		Molecular ion	Characteristic isotope peak at M + 2: 4.4* number of S atoms gives approx. intensity as percentage of peak at M.
		Fragments Rearrangements	$\frac{m}{e}$ = **M** − 48.

Table 16

Type of compound: **C, P, O compounds**

	Position	Assignment	Notes
IR	2440…2350 cm^{-1} m. 1300…1180 cm^{-1} s. ~ 1050 cm^{-1} s. 970… 910 cm^{-1} ~ 750 cm^{-1} s.	P—H st. P=O st. P—O—C st. as. P—O—P P—O—C st. sy.	 P=S st.: 800…650 cm^{-1} w. Broad bands.
UV	none above 210 nm		
NMR	~ 4 ppm ~ 7 ppm	CH_2—O—P H—P	$J_{CH—O—P}$ = 5…15Hz $J_{H—P}$ = 450…550Hz
MS		Molecular ion Fragments Rearrange-ments	Too little experimental data available.

Table 17

Type of compound: **N, O compounds**

	Position	Assignment	Notes
IR	1600…1500 cm^{-1} s. 1400…1300 cm^{-1} s.	{NO st. {NO_2 st. as. NO_2 st. sy.	Care: Table 25 cannot be used for aromatic nitro compounds.
UV			→ Table 28.
NMR			→ Table 41 or 43.
MS		Molecular ion Fragments Rearrange-ments	For Ar—NO_2: $\frac{m}{e} = 30, 46$; M − 46, M − 30.

Table 18

Type of compound: **C-halogen**

	Position	Assignment	Notes
IR	1400...1000 cm^{-1} s.	C—F st.	Ar—F st.: 1300...1200 cm^{-1}; CF_3: ~1350 cm^{-1}; ~1100 cm^{-1}.
	800...500 cm^{-1} s.	C—Cl st. C—Br st. C—I st.	C—Cl st.: ⩾ 600 cm^{-1} C—Br st.: ⩽ 600 cm^{-1} C—I st.: ~500 cm^{-1} Ar—Cl skel. *o*: 1060...1035 cm^{-1}. *m*: 1080...1075 cm^{-1}. *p*: 1095...1090 cm^{-1}.
UV	⩽ 280 nm (log ε = 2.5)	$n \rightarrow \sigma^*$	For C—Br and C—Cl usually only end absorption, none for C—F.
NMR	~2.6 ppm	CH_2I	
	~3.1 ppm	CH_2Br CH_2Cl CH_2F	See: $J_{H\text{-}C\text{-}F}$ = ~55Hz $J_{H\text{-}C\text{-}C\text{-}F}$ = 5...20 Hz (F-benzene, H at *o*, *m*, *p*) J_o = 8 Hz, J_m = 6 Hz, J_p = 1 Hz
MS		Molecular ion	Fluorides: present Chlorides:, Bromides: Characteristic isotope peaks, see Table 52. Iodides, polychlorides and -bromides: often missing.
		Fragments	Fluorides: $\frac{m}{e}$ = 19, 20; M − 20. Perfluorides: $\frac{m}{e}$ = 69 (CF_3), 119 (C_2F_5). Chlorides: $\frac{m}{e}$ = 35/37, 36/38, 49/51; M − 36/38. Iodides: $\frac{m}{e}$ = 127, 128; M − 127. Most important fragmentations: R—C—|—Hal > R—|—C—Hal
		Rearrangements	Chlorides and fluorides: M − H—Hal.

Table 19

Joint application of infrared, nuclear resonance, mass and electronic spectroscopy

Selection of literature

BRAND, J. C. D., & G. EGLINTON: Applications of Spectroscopy to Organic Chemistry. Oldbourne Press, London 1965

DIXON, R. N.: Spectroscopy and Structure. Methuen & Co. Ltd., London 1965.

DYER, J. R.: Applications of Absorption Spectroscopy of Organic Compounds. Prentice-Hall, Inc., Englewood Cliffs, N. J., 1965.

FREEMAN, S. K., (Ed.): Interpretive Spectroscopy. Reinhold Publishing Corp., New York 1965.

MATHIESON, D. W.: Interpretation of Organic Spectra. Academic Press, London, New York 1965.

PHILLIPS, J. P.: Spectra-Structure Correlation. Academic Press, New York, London 1964.

SCHWARZ, J. C. P., (Ed.): Physical Methods in Organic Chemistry. Oliver & Boyd, Edinburgh, London 1964.

SILVERSTEIN, R. M., & G. C. BASSLER: Spectrometric Identification of Organic Compounds. John Wiley and Sons, Inc., New York, London 1963.

YUKAWA, Y., (Ed.): Handbook of Organic Structural Analysis. W. A. Benjamin, Inc., New York, Amsterdam 1965.

D. H. WILLIAMS, I. FLEMING: Spectroscopic Methods in Organic Chemistry. McGraw-Hill Publishing Co. Ltd., London, New York, Toronto, Sydney, 1966.

Table 20

Prohibited regions of some of the solvents and suspension media commonly used in infrared spectroscopy

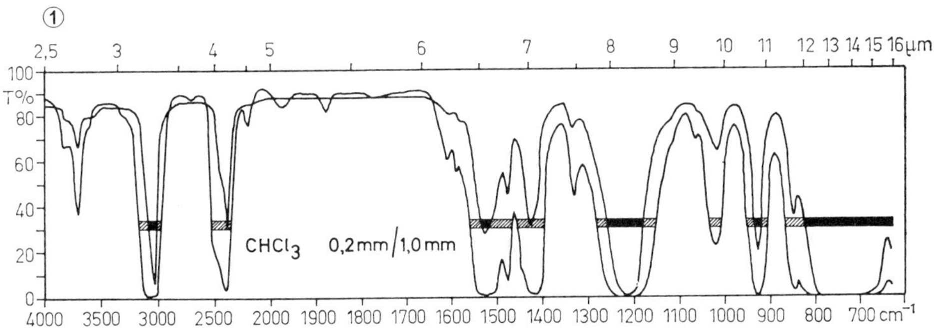

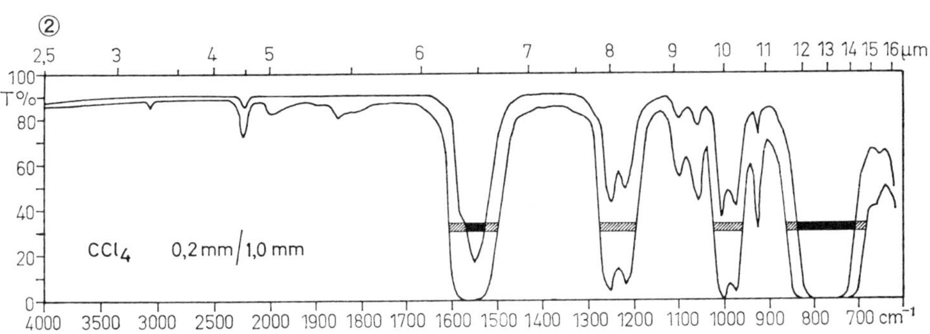

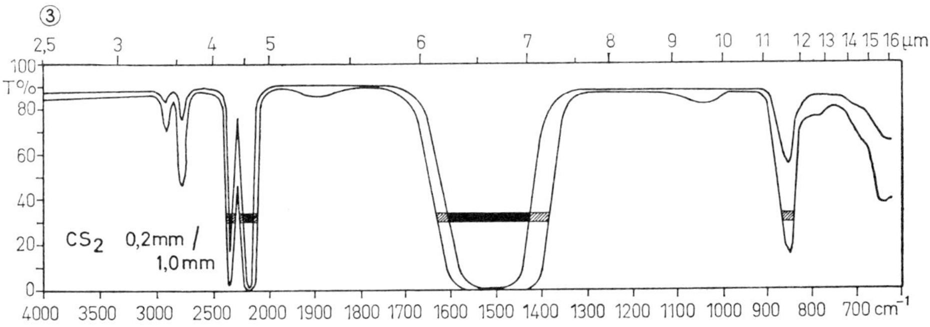

Continuation of Table 20

Prohibited regions of some of the solvents and suspension media commonly used in infrared spectroscopy

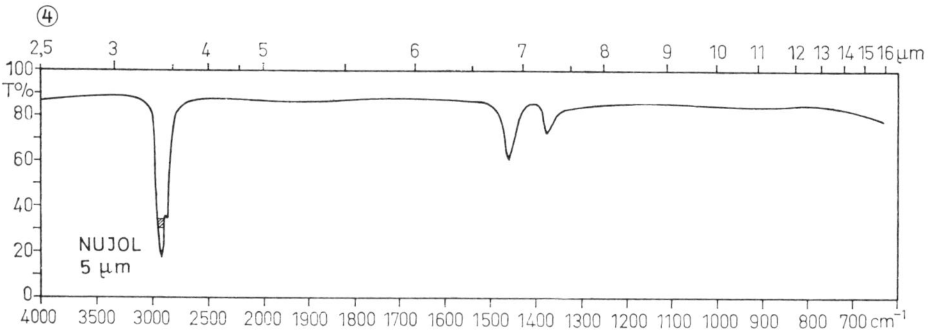

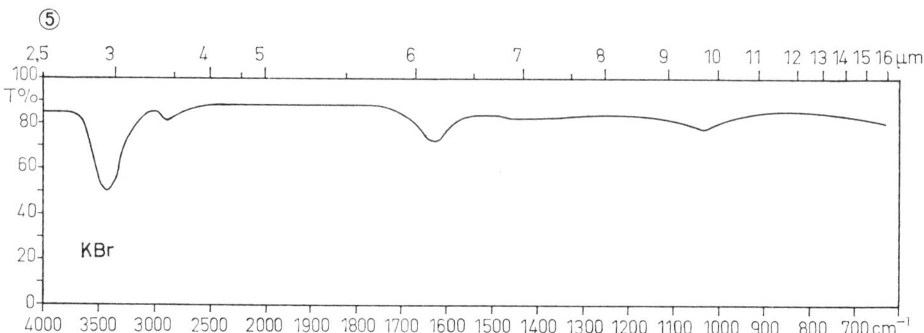

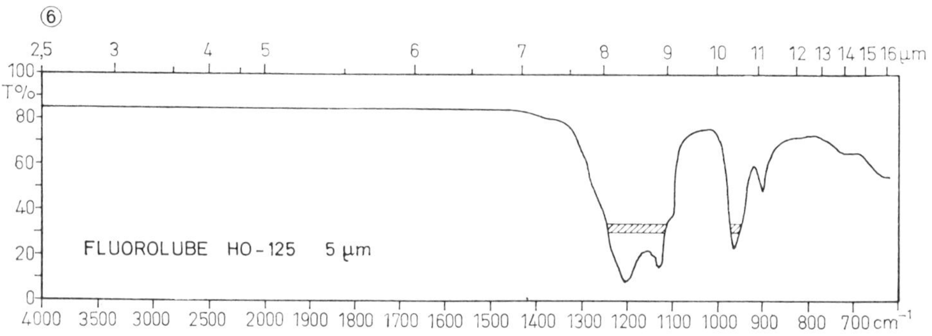

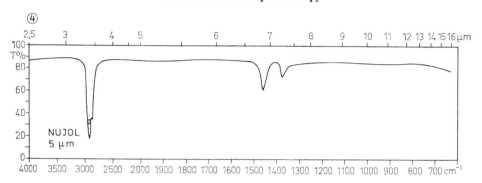

Table 21

C—H deformation vibrations (cm^{-1})

X	$X—CH_3$		$X—CH_2$
	def. as.	def. sy.	def.
>C< (—C— with bonds above and below)	1475...1445	1390...1370	1475...1445
—O—	1470...1450	1460...1440	1475...1445
—N<	1460	1440...1410	1475...1445
—C(=O)—	1420	1375...1345	1445...1405
C=C	1460	1380	1455...1453

Table 22

X≡Y and X=Y=Z stretching vibrations

Type of compound	**Position** (cm^{-1})
—C≡C—	2260...2190
—C≡C—H	2140...2100
—C=C=C	1975...1910
—C≡N	2260...2210
—N=C=O	2275...2250
—N=C=S	2140...1990
—S—C≡N	2175...2120
$—N_3$	2160...2120

Table 23

Skeletal vibrations of compounds containing $-C(CH_3)_2-$

(Band at 1380 cm^{-1} split)

Type of compound	**Position** (cm^{-1})	Notes
$H-C(CH_3)_2-C$	1170 s. 1150	 Shoulder
$CH_3-C(CH_3)_2-C$	1255 s. 1250...1200 s. 925	
$C-C(CH_3)_2-C$	1215 1195 s.	Shoulder

Table 24

Infrared absorption of compounds of types C=C (cm^{-1})

Type of compound	**Assignment**						
	=CH st.	Overtone of CH def. (out-of-plane)	C=C st.	CH def. (out-of-plane)			
				C=C isol.	C=C—C=O	C=C—OR	C=C—O—C(=O)—R
CHR=CH_2	3095...3075 m. 3040...3010 m.	1850...1800 m.	1645...1640 v.	1005...985 m. 920...900 s.	~980 ~960	~960 ~820	~950 ~870
CHR=CHR cis.	3040...3010 m.		1665...1635 v.	730...665 s.	~820		
CHR=CHR trans	3040...3010 m.		1675...1665 v.	990...960 s.	~975	~960	~950
CR_2=CH_2	3095...3075 m.	1800...1780 m.	1660...1640 v.	900...880 s.	~940	~800	
CR_2=CHR	3040...3010 m.		1690...1670	850...790 m.			
CR_2=CR_2	—	—	1690...1670	—	—	—	—
Ar—C=C			1635...1615				
O=C—C=C			1665...1585				
$(C=C)_{2...3}$			1650 1600				
$(C=C)_{n\geq 4}$			1650...1580 broad				

Table 25

Infrared absorption of aromatic compounds in cm^{-1}

Substitution type

	Mono	*o*-Di	*m*-Di	*p*-Di	1,2,3-Tri	1,2,4-Tri
ArH def. (in-plane) often some bands missing, no practical importance	1170...1125 w. 1110...1070 w. 1070...1000 w.	1225...1175 w. 1125...1090 w. 1070...1000 w. 1070... 960 w.	1175...1125 w. 1110...1070 w. 1070...1000 w.	1225...1175 w. 1125...1090 w. 1070...1000 w. 1070...1000 w.	1175...1125 w. 1110...1070 w. 1070...1000 w. 1000... 960 w.	1225...1175 w. 1175...1125 w. 1125...1090 w. 1070...1000 w. 1070...1000 w. 1000... 960 w.
ArH def. (out-of-plane) Ring def.	910...890 s. (sometimes missing) 770... 730 s. 710... 690 s.	770... 735 s.	900... 860 m. 810... 750 s. 725...680 m. (sometimes missing)	860... 800 s.	810... 750 s. 725...680 m. (sometimes missing)	900... 860 m. 860... 800 s.
Combination vibrations in region 2000...1660 cm^{-1}		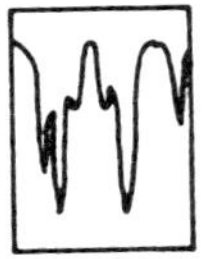				

	1,3,5-Tri	1,2,3,4-Tetra	1,2,3,5-Tetra	1,2,4,5-Tetra	Penta	Hexa
ArH def. (in-plane) often some bands missing, no practical importance	1175...1125 w. 1070...1000 w.					
ArH def. (out-of-plane) Ring def.	900... 860 m. 865... 810 s. 730... 690 s.	820... 800 s.	860... 840 m.	900...850 m.	900... 860 m.	—
Combination vibrations in region 2000...1660 cm^{-1}	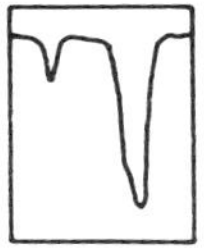	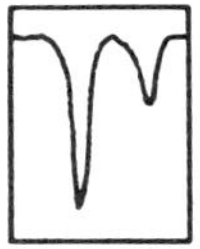	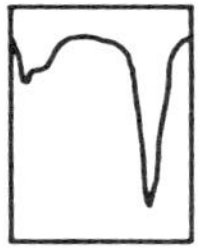	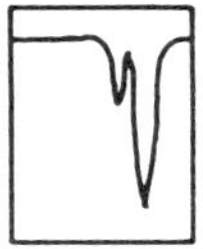		

Infrared absorption of aromatic compounds in cm^{-1}

Also of use for heterocycles such as pyridine, quinoline etc.; the hetero-atom is considered as a substituent; in general the following are values for ArH def. (out-of-plane):

5 neighbouring H atoms:	770...730 cm^{-1}
4 neighbouring H atoms:	770...735 cm^{-1}
3 neighbouring H atoms:	810...750 cm^{-1}
2 neighbouring H atoms:	860...800 cm^{-1}
1 isolated H atom:	900...860 cm^{-1}

For compounds of type

the aromatic proton signals in NMR spectrum show a plane of symmetry.

Caution: If substituents are present which conjugate strongly with the ring system (carbonyl, nitro groups) the ranges given are no longer valid.

Infrared spectroscopy

Selection of literature

Textbooks and monographs

BELLAMY, L. J.: The Infra-red Spectra of Complex Molecules. Methuen & Co., Ltd., London, and John Wiley & Sons, Inc., New York, 1960.

BIBLE, R. H., jnr.: Guide to the NMR Empirical Method. Plenum Press, New York, 1965.

BRÜGEL, W.: Einführung in die Ultrarotspektroskopie. Dr. Dietrich Steinkopff Verlag, Darmstadt 1962.

*COLTHUP, N. B., L. H. DALY u. S. E. WIBERLEY Introduction to Infrared and Raman Spectroscopy. Academic Press, New York London, 1964.

CROSS, A. D.: Introduction to Practical Infra-Red Spectroscopy. Butterworths Scientific Publications, London 1960.

HENNIKER, J. C.: Infrared Spectrometry of Industrial Polymers. Academic Press, London, New York, 1967.

*NAKANISHI, K.: Infrared Absorption Spectroscopy, Practical. Holden-Day, Inc., San Francisco and Nankodo Company Ltd., Tokyo 1962.

OTTING, W.: Spektrale Zuordnungstafel der Infrarot-Absorptionsbanden. Springer-Verlag, Berlin, Göttingen, Heidelberg 1963.

SZYMANSKI, H. A.: Interpreted Infrared Spectra, Volume I und II. Plenum Press, New York 1964 und 1966.

---

* Especially to be recommended for interpretation of spectra of organic compounds.

Table 27

Infrared spectroscopy

Selection of literature

Collection of data

HERSHENSON, H. M.: Infrared Absorption Spectra, Index for 1958 to 1962. Academic Press, New York, London, 1964.

HERSHENSON, H. M.: Infrared Absorption Spectra, Index for 1945 to 1957. Academic Press, New York, London, 1964.

Dokumentation der Molekülspektroskopie (DMS-Kartei). Institut für Spektrochemie und angewandte Spektroskopie, Dortmund. Verlag Chemie GmbH, Weinheim.

American Society for Testing Materials. Wyandotte Punched Card Index to Infrared Absorption Spectra, ASTM, Philadelphia.

Sadtler Standard Spectra. Sadtler Research Laboratories, Philadelphia.

American Petroleum Institute Research Project 44, Carnegie Institute of Technology, Pittsburgh.

Standard Infrared Spectral Data Cards (IRDC Cards [Japan]), The Infrared Data Committee of Japan, Infrared and Raman Discussion group. Tokyo.

OTTING, W.: Spektrale Zuordnungstafel der Infrarot-Absorptionsbanden. Springer-Verlag, Berlin, Göttingen, Heidelberg, 1963.

SZYMANSKI, H. A.: Interpreted Infrared Spectra, Volume I, Volume II, Volume III. Plenum Press, New York, 1964, 1966, 1967.

Table 28

UV absorption of simple chromophores

Chromophore	Compound	Transition	λ_{max}[nm]	log ε	Solvent
$C=C$	$CH_2=CH_2$	$\pi \rightarrow \pi^*$	162.5	4.2	heptane
	$(CH_3)_2C=C(CH_3)_2$	$\pi \rightarrow \pi^*$	196.5	4.1	heptane
$C=O$	$(CH_3)_2C=O$	$n \rightarrow \pi^*$	279	1.2	cyclohexane
		$\pi \rightarrow \pi^*$	188	3.3	cyclohexane
	$CH_3C(=O)OH$	$n \rightarrow \pi^*$	197	1.8	hexane
$C=N$	$(CH_3)_2C=N—OH$		193	3.3	ethanol
$C=S$	$C_2H_5OC(=S)—CH_3$		241	2.9	ethanol
			369	1.3	
$N=N$	$CH_3—N=N—CH_3$	$n \rightarrow \pi^*$	345	0.7	ethanol
$N=O$	$(CH_3)_3C—N=O$		300	2.0	ethyl ether
			665	1.3	
NO_2	CH_3NO_2		278	1.3	ethyl ether
$C\equiv C$	$H—C\equiv C—H$	$\pi \rightarrow \pi^*$	173	3.8	gas phase
$C\equiv N$	$CH_3—C\equiv N$		<190		liquid
$C—H$	CH_4	$\sigma \rightarrow \sigma^*$	122	high	gas phase
$C—C$	CH_3CH_3	$\sigma \rightarrow \sigma^*$	135	high	gas phase
$C—O$	CH_3OH	$n \rightarrow \sigma^*$	177	2.3	hexane
	CH_3OCH_3		<185		hexane
$C—Cl$	CH_3Cl	$n \rightarrow \sigma^*$	173	2.3	hexane
$C—Br$	C_3H_7Br	$n \rightarrow \sigma^*$	208	2.5	hexane
$C—I$	CH_3I	$n \rightarrow \sigma^*$	259	2.6	hexane
$C—S$	$CH_3CH_2—S—CH_2CH_3$		194	3.7	hexane
			215	3.2	
$S—S$	$CH_3CH_2—S—S—CH_2CH_3$		194	3.7	hexane
			250	2.6	

Table 29

UV-Absorption of dienes and polyenes*

(Woodward rule for $\pi \rightarrow \pi^*$ transition)

Heteroannular dienes or polyenes of type

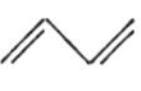  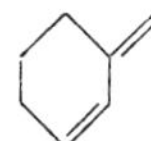 

— Base value for heteroannular diene:		214 nm

Homoannular dienes or polyenes of type

— Base value for homoannular diene:		253 nm
— Increment: for C substituent:		+ 5 nm
	for exocyclic double bond	+ 5 nm
	for double bond extending conjugation	+ 30 nm
	for polar group: —OAc:	0 nm
	—O Alkyl:	+ 6 nm
	—S Alkyl:	+ 30 nm
	—Cl, —Br:	+ 5 nm
	—N(Alkyl)$_2$:	+ 60 nm
— Correction for solvent		0 nm

— In the case of crossed conjugated systems, the value is calculated for the chromophore absorbing at the longest wavelength.

* R. B. Woodward: J. Amer. Chem. Soc. **63**, 1123 (1941); **64**, 72, 76 (1942); L. F. Fieser, M. Fieser: Natural Products Related to Phenanthrene, S. 184—198, Reinhold Publishing Company, New York 1949; A. I. Scott: Interpretation of the Ultraviolet Spectra of Natural Products, Pergamon Press, Oxford, London, Edinburgh, New York, Paris, Frankfurt 1964.

UV absorption of dienes and polyenes*

Examples:

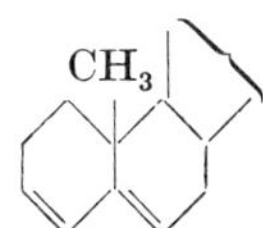

Calculated: 234 nm = (214 + 3 × 5 + 5) nm
Found: 234 nm (log ε = 4.3) (ether) for cholesta-3,5-diene

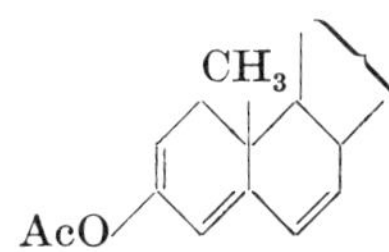

Calculated: 303 nm = (253 + 3 × 5 + 5 + 30) nm
Found: 304 nm (log ε = 4.2) (ethanol) for ergost-4,6,22-trien-3-one enol acetate

Table 30

UV absorption spectra

(in ethanol)

Chromophore: **C=C—C=C** (see also Table 29)

Compound	λ_{max} [nm]	log ε
$CH_2=CH-CH=CH_2$	217	4.3
$CH_2=CR-CH=CH_2$	220	4.3
$RCH=CH-CH=CH_2$	223	4.4
$CH_2=CR-CR=CH_2$	226	4.3
$RCH=CH-CH=CHR$	227	4.4
⬡$=CH-CH=CH_2$	237	3.9
⬡$=CH-CH=$⬡	247	4.3
$CH_3-(CH=CH)_3-CH_3$ trans	275	4.5
$CH_3-(CH=CH)_4-CH_3$ trans	310	4.9
$CH_3-(CH=CH)_5-CH_3$ trans	341	5.1

Table 31

UV absorption spectra

(in ethanol)

Chromophore: **C=C—C=O** (see also Table 32)

Compound	λ_{max} [nm] $\pi\rightarrow\pi^*$	log ε	λ_{max} [nm] $n\rightarrow\pi^*$	log ε
$CH_2=CH—CHO$	208	4.0	328	1.1
$CH_3—CH=CH—CHO$	220	4.2	322	1.4
$CH_3—(CH=CH)_2—CHO$	271	4.4	not visible	
$CH_2=CH—CO—CH_3$	212	3.5	324	1.3
$CH_2=C(CH_3)—CO—CH_3$	218	3.9	319	1.4
$CH_3—CH=CH—CO—CH_3$	224	4.0	315	1.6
$(CH_3)_2C=CH—CO—CH_3$	235	4.2	314	1.8
Measurement in hydrocarbon	shorter		longer	

Table 32

UV absorption spectra

(in ethanol)

Chromophore: **C=C—COOR**

Compound	λ_{max} [nm]	log ε
CH_2=CH—COOH	<200	—
CH_3—CH=CH—COOH	205	4.2
$(CH_3)_2C$=CH—COOH	216	4.1
$(CH_3)_2C$=C(CH_3)—COOH	221	4.0
CH_3—$(CH{=}CH)_2$—COOH	254	4.4
CH_3—$(CH{=}CH)_3$—COOH	294	4.6
CH_3—$(CH{=}CH)_4$—COOH	327	4.7
CH_2=CH—COOR	<200	—
CH_3—CH=CH—COOR	205	4.2
$(CH_3)_2C$=CH—COOR	217	4.2
ROOC—CH=CH—COOR cis	205	3.9
ROOC—CH=CH—COOR trans	211	4.2

Table 33

UV absorption of α,β-unsaturated ketones*

(Woodward rule for $\pi \rightarrow \pi^*$ transition)

$$\begin{array}{c} \beta \\ \beta \end{array}\!\!>\!C{=}\overset{\alpha}{C}{-}\overset{R}{C}{=}O \qquad\qquad \begin{array}{c} \delta \\ \delta \end{array}\!\!>\!C{=}\overset{\gamma}{C}{-}\overset{\beta}{C}{=}\overset{\alpha}{C}{-}\overset{R}{C}{=}O$$

I II

– Basic system I:

double bond in six-membered ring or acyclic:		215 nm
double bond in five-membered ring:		202 nm

– Basic system II: six-membered ring or acyclic: 245 nm

– Increment:

for C substituent in	α:	+ 10 nm
	β:	+ 12 nm
	γ:	+ 18 nm
	δ (and further)	+ 18 nm
for exocyclic double bond:		+ 5 nm
for conjugated double bond:		+ 30 nm
for homodiene component:		+ 39 nm
for polar group: —OH	α:	+ 35 nm
	β:	+ 30 nm
	δ:	+ 50 nm
—OAc	α, β, δ:	+ 6 nm
—OMe	α:	+ 35 nm
	β:	+ 30 nm
	γ:	+ 17 nm
	δ:	+ 31 nm
—S Alkyl	β:	+ 85 nm
—Cl	α:	+ 15 nm
	β:	+ 12 nm
—Br	α:	+ 25 nm
	β:	+ 30 nm
—N $(Alkyl)_2$	β:	+ 95 nm

* R. B. Woodward: J. Amer. Chem. Soc. **63**, 1123 (1941); **64**, 72, 76 (1942); L. F. Fieser, M. Fieser: Natural Products Related to Phenanthrene, S. 184—198, Reinhold Publishing Company, New York 1949; A. I. Scott: Interpretation of the Ultraviolet Spectra of Natural Products, Pergamon Press, Oxford, London, Edinburgh, New York, Paris, Frankfurt 1964.

Table 33 (cont.)

UV absorption of α,β-unsaturated ketones*

— Solvent correction:	ethanol	0
	methanol	0
	dioxan	— 5 nm
	chloroform	— 1 nm
	ether	— 7 nm
	water	+ 8 nm
	hexane	— 11 nm
	cyclohexane	— 11 nm

— In the case of crossed conjugated systems the value must be calculated for the chromophore absorbing at the longest wavelength.

Examples:

Calculated: 280 nm = (215+30+12+18+5) nm

Found: 283 nm (log ε = 4.3) (ethanol) for 6-dehydrocorticosterone

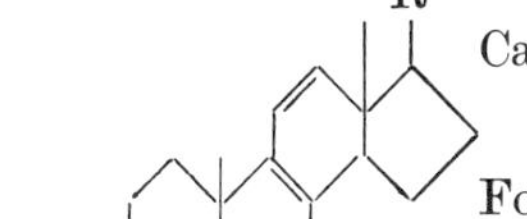

Calculated: 324 nm = (215+30+10+12+18+39) nm

Found: 256 nm and 327 nm (ethanol)

Table 34

UV absorption spectra

Chromophore: **aromatic**

Compound	Solvent	λ_{max} (nm) (log ε) of band at longest wavelength
benzene	hexane	204 (3.9); 254 f. (2.4)
$C_6H_5CH_3$	hexane	208 (3.9); 262 f. (2.4)
C_6H_5Cl	heptane	216 (3.9); 265 f. (2.4)
$C_6H_5NO_2$	hexane	251 (3.9); 322 (2.2)
C_6H_5OH	water	211 (3.8); 270 (3.2)
$C_6H_5O^{\ominus}$	water	235 (4.0); 287 (3.4)
$C_6H_5OCH_3$	water	220 (3.9); 272 (3.2)
C_6H_5COOH	water	226 (4.0); 271 (2.9)
$C_6H_5COO^{\ominus}$	water	223 (4.0); 267 (2.8)
$C_6H_5COOCH_3$	water	230 (4.0); 273 (3.0)
$C_6H_5NH_2$	water	230 (3.9); 281 (3.1)

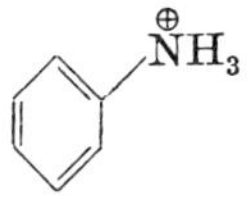 $\overset{\oplus}{N}H_3$	water	203 (3.9); 254 f. (2.2)
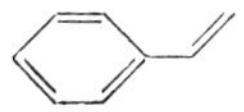	heptane	248 (4.2); 273 (2.9); 282 (2.9); 291 (2.8)
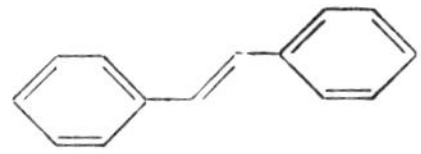	ethanol	228 (4.2); 295 (4.5); 307 (4.5)
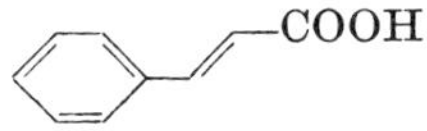 COOH	water	216 (4.2); 270 (4.2)
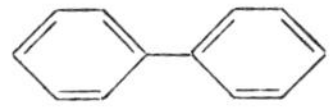	heptane	201 (4.6); 247 (4.2)
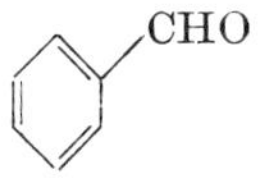 CHO	ethanol	244 (4.0); 280 (3.2)
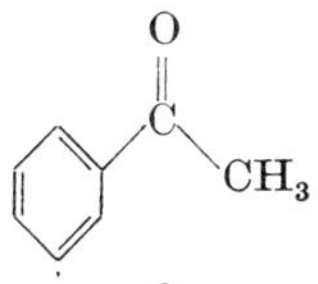 O, C, CH_3	ethanol	242 (4.1); 279 (3.0); 318 s. (1.8)
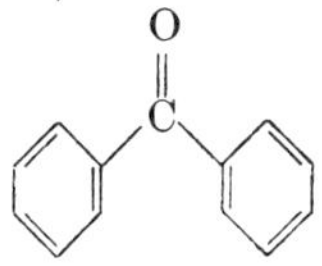 O, C	ethanol	252 (4.2); 333 (2.2)
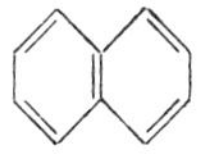	hexane	221 (5.0); 275 f. (3.7); 305 f. (2.3)
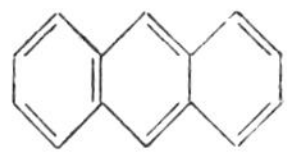	hexane	221 (4.0); 252 (5.1); ~360 f. (2.4)
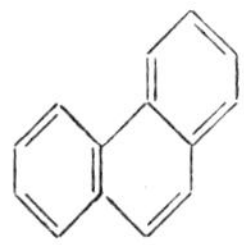	hexane	211 (4.5); 251 (4.8); 292 (4.1); ~330 f. (2.4)

Table 35

UV absorption spectra
(in ethanol)
Chromophore: **heterocycle**

Compound	λ_{max}[nm] (log ε)
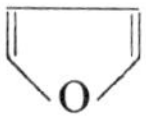 Furan	<210
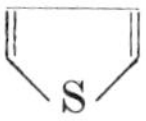 Thiophen	215 (3.8); 231 (3.9)
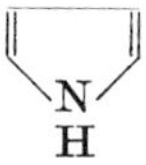 Pyrrole	<220
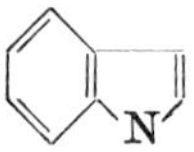 Indole	218 (4.4); 271 (3.8); 278 (3.8); 287 (3.7)
 Pyridine	198 (3.7); ~255 f. (3.3)
 Pyridazine	192 (3.7); ~245 f. (3.2); 311 (2.5)
 Pyrimidine	243 (3.5); 280 (2.6)
 Pyrazine	~260 f. (3.7); ~315 f. (3.5)
 Quinoline	203 (4.7); 228 (4.4); 232 (4.4); ~275 f. (3.5)
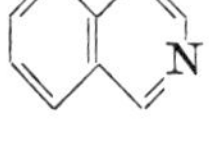 Isoquinoline	217 (4.7); 258 (3.5); 280 (3.3); ~305 f. (3.3)
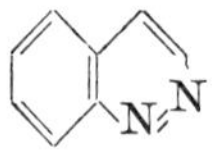 Cinnoline*	222 (4.6); 276 (3.4); 308 (3.2); 322 (3.3); 390 (2.4)
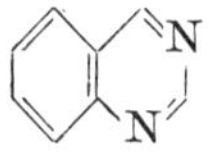 Quinazoline*	220 (4.6); 267 (3.4); 311 (3.2)

Table 35 (cont.)

Compound	λ_{max}[nm] (log ε)
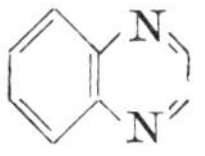 Quinoxaline*	233 (4.5); 316 f. (3.8); 355 (2.8)
 Phthalazine*	215 (4.7); 259 (3.7); 303 (3.0)
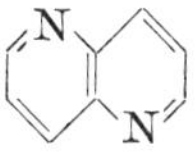 1,5-Naphthyridine*	235 (3.8); 295 (3.5)
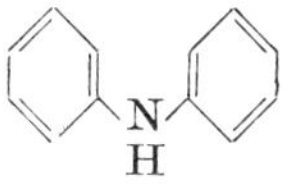 Carbazole	234 (4.6); 245 (4.4); 257 (4.3); 293 (4.2); 324 (3.5); 337 (3.5)
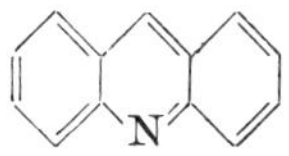 Acridine	250 (5.4); 351 f. (4.0)
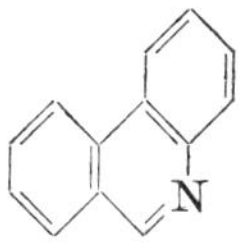 Phenanthridine	245 (4.7); 290 (3.9); 330 (3.3); 345 (3.3)
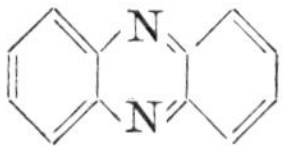 Phenazine	250 (5.3); 350 s. (4.0); 360 (4.2); 390 s. (3.5)

s.: shoulder

* in cyclohexane

Table 36

UV absorption of aromatic compounds*

(charge transfer bands)

Solvent: ethanol

R		R-C_6H_5 (benzene ring)		
Basic system:		*Increment for each substituent:*		
—CO-alkyl	246 nm	—Alkyl or ring	o-, m-	3 nm
—CO-Ring	246 nm		p-	10 nm
—COH	250 nm	—OH, —OMe, —OAlk	o-, m-	7 nm
—COOH	230 nm		p-	25 nm
—COO-Alkyl	230 nm	$—O^-$	o-	11 nm
—COO-Ring	230 nm		m-	20 nm
—CN	224 nm		p-	78 nm
		—Cl	o-, m-	0 nm
			p-	10 nm
		—Br	o-, m-	2 nm
			p-	15 nm
		$—NH_2$	o-, m-	13 nm
			p-	58 nm
		—NHAc	o-, m-	20 nm
			p-	45 nm
		—NHMe	p-	73 nm
		$—NHMe_2$	o-, m-	20 nm
			p-	85 nm

Table 37

Spectroscopy in UV and visible regions

Selection of literature

Textbooks and monographs

Gillam, A. E., and E. S. Stern: An Introduction to Electronic Absorption Spectroscopy in Organic Chemistry. Edward Arnold (Publishers) Ltd., London 1957.

* A. I. Scott: Interpretation of the Ultraviolet Spectra of Natural Products. Pergamon Press, Oxford, London, Edinburgh, New York, Paris, Frankfurt 1964.

Jaffé, H. H., and M. Orchin: Theory and Application of Ultraviolet Spectroscopy. J. Wiley & Sons, Inc., New York 1966.

Murrell, J. N.: The Theory of the Electronic Spectra of Organic Molecules. Methuen & Co., Ltd., London, J. Wiley & Sons, Inc., New York 1964.

Scott, A. I.: Interpretation of the Ultraviolet Spectra of Natural Products. Pergamon Press, Oxford, London, Edinburgh, New York, Paris, Frankfurt 1964.

Beaven, G. H., E. A. Johnson, H. A. Willis, and R. G. J. Miller: Molecular Spectroscopy. Heywood & Co. Ltd., London 1961.

Rao, C. N. R.: Ultraviolet and Visible Spectroscopy. Plenum Press, New York, Washington, London 1967.

Suzuki, H.: Electronic Absorption Spectroscopy of Organic Molecules. Academic Press, New York 1967.

Table 38

Spectroscopy in UV and visible regions

Selection of literature

Collection of data

Friedel, R. A., and M. Orchin: Ultraviolet Spectra of Aromatic Compounds. J. Wiley & Sons Inc., New York 1951. Revised ed. 1958.

Lang, L.: Absorption Spectra in the Ultraviolet and Visible Region. Academic Press, New York 1961.

Kamlet, M. J., and H. E. Ungnade (eds.): Organic Electronic Spectral Data, Volume I: 1946–1952; Volume II: 1953–1955. Interscience, New York 1960.

Hershenson, H. M.: Ultraviolet Absorption Spectra, Index for 1930–1954, 1955–1959, 1960–1963. Academic Press, New York 1956, 1961, 1966.

Hirayama, K.: Handbook of Ultraviolet and Visible Absorption Spectra of Organic Compounds. Plenum Press, New York, Washington, London 1967.

Organic Electronic Spectral Data, Volume I–IV, 1946–1959. Interscience, New York 1960–1966.

Catalog of Ultraviolet Absorption Spectrograms. American Petroleum Institute, Project 44, Carnegie Institute of Technology, Pittsburgh.

Sadtler Ultraviolet Spectra, Collection of 11 000 Ultraviolet Spectra. Sadtler Research Laboratories, Philadelphia, Pa., Heyden & Son, Ltd., London.

UV-Atlas organischer Verbindungen (DMS). Verlag Chemie, Weinheim und Butterworths, London 1966.

Table 39

NMR spectra of common solvents and reference compounds

Recorded at 60 MHz

Conversion of chemical shifts based on various internal reference compounds X(δ_X ppm) into shifts based on TMS as internal standard:

$$\delta_{TMS} = \delta_X + K_X$$

X	K_X
cyclohexane	1.43
acetone	2.09
DMSO	2.55
dioxan	3.57
water	4.79
methylene chloride	5.33
chloroform	7.25
benzene	7.27

Spectrum No. 1. 2,2-Dimethyl-2-silapentan-5-sulphonate (DSS)
$(CH_3)_3Si(CH_2)_3SO_3Na$ in D_2O

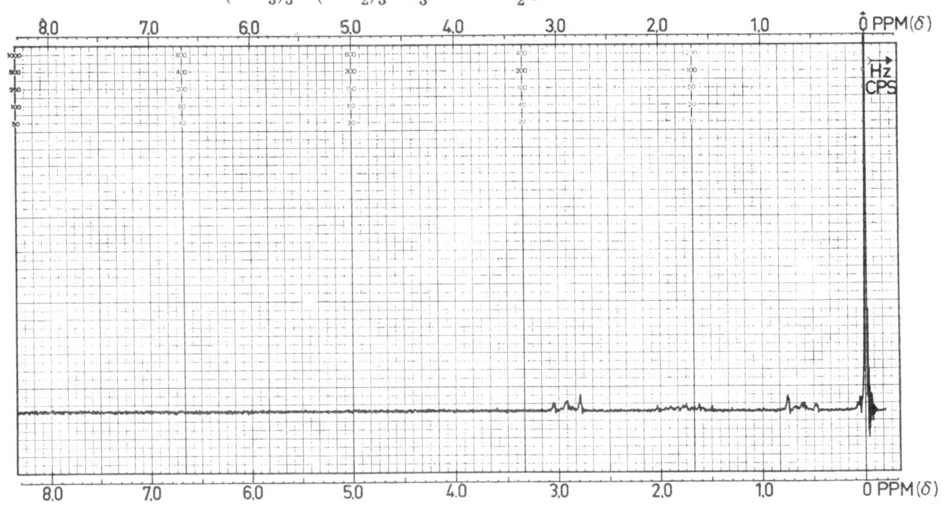

Table 39 (cont.)

NMR spectra of common solvents and reference compounds

Recorded at 60 MHz

Spectrum No. 2. Deuterochloroform $CDCl_3$

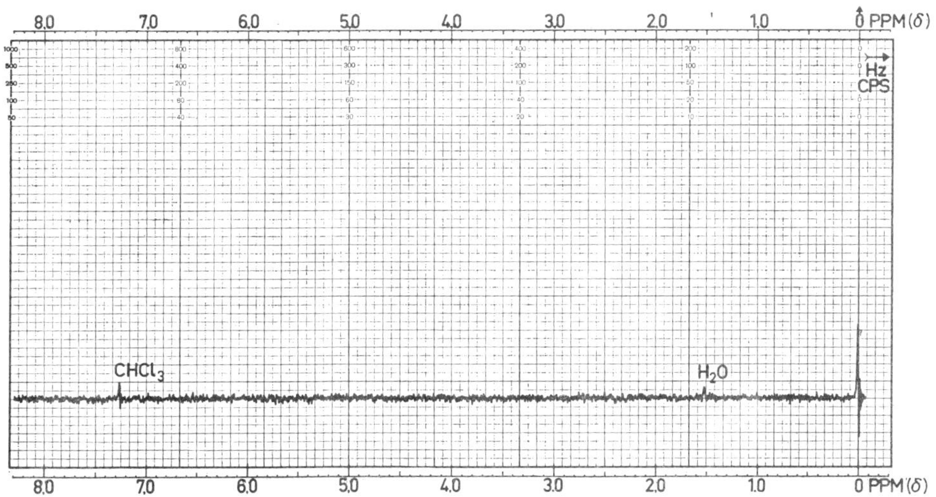

Spectrum No. 3. Deuterium oxide D_2O

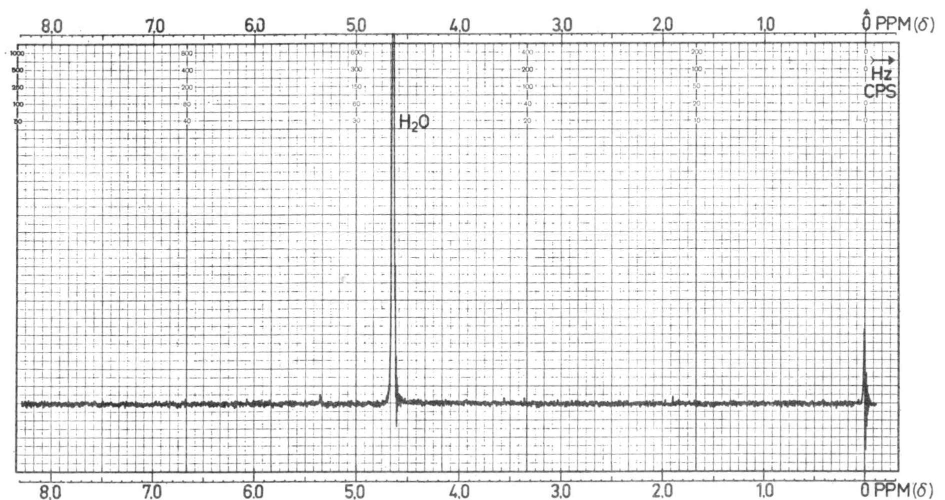

Table 39 (cont.)

NMR spectra of common solvents and reference compounds

Recorded at 60 MHz

Spectrum No. 4. Trideuteroacetonitrile, acetonitrile d_3, CD_3CN

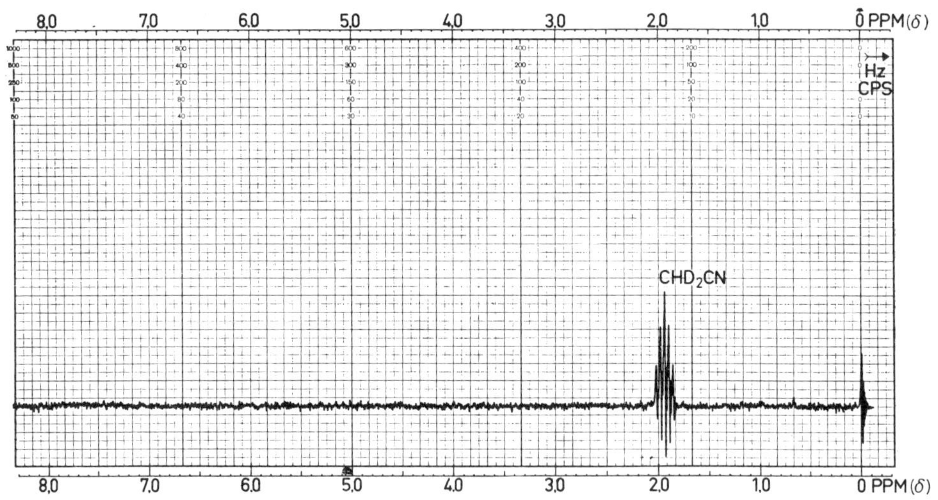

Spectra No. 5. Pentadeuteropyridine, pyridine d_5, C_5D_5N

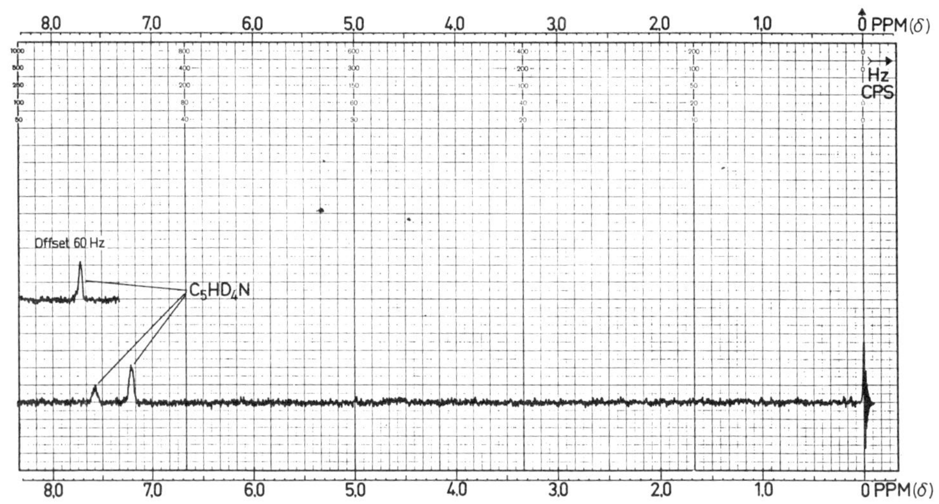

Table 39 (cont.)

NMR spectra of common solvents and reference compounds

Recorded at 60 MHz

Spectra No. 6. Hexadeuteroacetone, acetone d_6, CD_3COCD_3

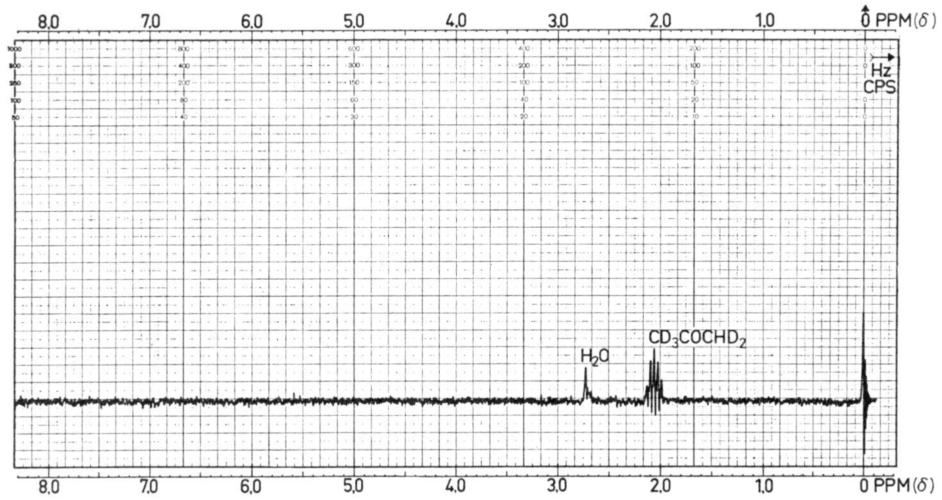

Spectra No. 7. Hexadeuterodimethyl sulphoxide, dimethyl sulphoxide d_6, CD_3SOCD_3

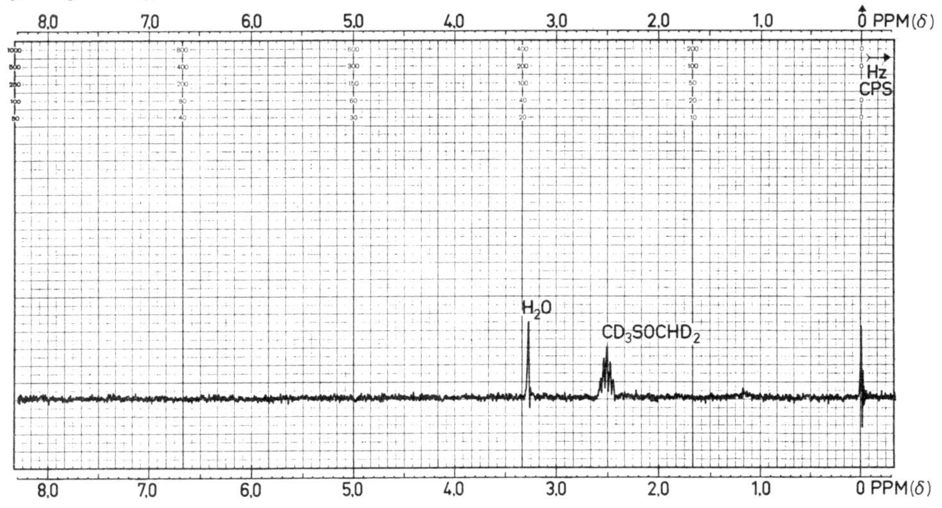

Table 40

Chemical shifts δ of protons linked in various ways

in ppm based on tetramethylsilane. The ranges shown only hold if no additional electronegative substituents are adjacent.

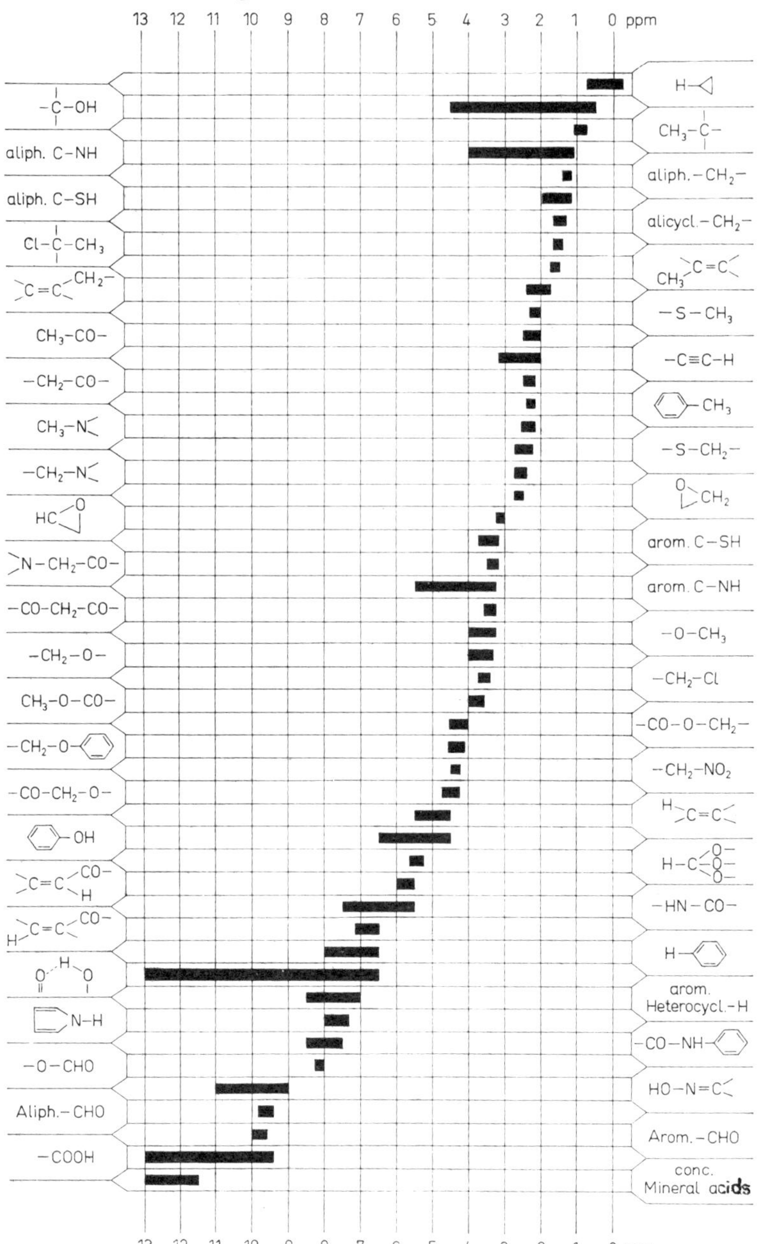

Table 41

Estimation of chemical shift δ for protons of
—CH_2— and >CH—groups

(in ppm, based on tetramethylsilane)
(modified Shoolery's rules)

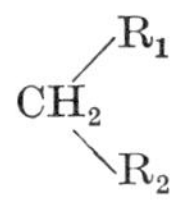

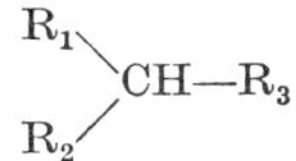

$$\delta_{CH_2} = 1.25 + \sum_i a_i$$

$$\delta_{CH} = 1.5 + \sum_i a_i$$

(Estimation less certain)

R_i	a_i	R_i	a_i
—Alkyl	0.00	—C_6H_5 (phenyl)	1.3
—C=C	0.75	—Br	1.9
—C≡C	0.90	—Cl	2.0
—COOH, —COOR	0.7	—OR, —OH	1.7
—CO—R, —CN	1.2	—O—CO—R	2.7
—S—R	1.0	—O—C_6H_5 (phenyl)	2.3
—NH_2, —NR_2	1.0	—NO_2	3.0
—I	1.4		

Table 42

Estimation of chemical shift δ for protons at a double bond*

The chemical shifts of an olefinic proton ($\delta_{C=C\backslash H}$) in ppm from internal tetramethylsilane are given by:

$$\delta_{C=C\backslash H} = 5.25 + Z_{gem} + Z_{cis} + Z_{trans}$$

$R_{cis}(R_{trans})C{=}C(H)(R_{gem})$

Substituent R	Z_i for R (ppm)		
	Z_{gem}	Z_{cis}	Z_{trans}
1. —H	0	0	0
2. —Alkyl	0.45	−0.22	−0.28
3. —Alkyl-Ring	0.69	−0.25	−0.28
4. —CH_2O	0.64	−0.01	−0.02
5. —CH_2S	0.71	−0.13	−0.22
6. —CH_2X, X: F, Cl, Br	0.70	0.11	−0.04
7. —CH_2N	0.58	−0.10	−0.08
8. —C=C isolated	1.00	−0.09	−0.23
9. —C=C conjugated	1.24	0.02	−0.05
10. —C≡N	0.27	0.75	0.55
11. —C≡C	0.47	0.38	0.12
12. —C=O isolated	1.10	1.12	0.87
13. —C=O conjugated	1.06	0.91	0.74
14. —COOH isolated	0.97	1.41	0.71
15. —COOH conjugated	0.80	0.98	0.32
16. —COOR isolated	0.80	1.18	0.55
17. —CR conjugated	0.78	1.01	0.46
18. —C(H)=O	1.02	0.95	1.17
19. —C(N)=O	1.37	0.98	0.46
20. —C(Cl)=O	1.11	1.46	1.01

* Matter, U. E., C. Pascual, E. Pretsch, A. Pross, W. Simon and S. Sternhell: Tetrahedron, **25**, 691 (1969).

Substituent R	Z_i for R (ppm)		
	Z_{gem}	Z_{cis}	Z_{trans}
21. —OR, R: aliphatic	1.22	—1.07	—1.21
22. —OR, R: conjugated	1.21	—0.60	—1.00
23. —OCOR	2.11	—0.35	—0.64
24. $—CH_2—C{=}O$; $—CH_2—C{\equiv}N$	0.69	—0.08	—0.06
25. $—CH_2$-Aromatic-Ring	1.05	—0.29	—0.32
26. —Cl	1.08	0.18	0.13
27. —Br	1.07	0.45	0.55
†28. —I	1.14	0.81	0.88
29. —N—R, R: aliphatic	0.80	—1.26	—1.21
30. —N—R, R: conjugated	1.17	—0.53	—0.99
31. —N—C=O	2.08	—0.57	—0.72
32. —Aromatic	1.38	0.36	—0.07
33. —Aromatic fixed	1.60	—	—0.05
34. —Aromatic *o*-subst.	1.65	0.19	0.09
35. —SR	1.11	—0.29	—0.13
36. $—SO_2$	1.55	1.16	0.93

* The values in this row are based on data from only 4 substances.

Comment on Table 42. The increments 'R conjugated' are to be used instead of 'R isolated' when either the substituent or the double bond is conjugated with further substituents. The increments 'Alkyl-Ring' are to be used when the substituent together with the double bond are a part of a cyclic structure.* The increments 'Aromatic fixed' are to be used when the double bond is conjugated with the aromatic ring and is also a part of a ring closed to the aromatic ring (e.g. 1,2-dihydronaphthalene).

* Data for compounds containing 3- and 4-membered rings have not been considered.

Table 43

Influence of a substituent on the chemical shift of ring protons in monosubstituted benzene

$\delta_{Benzene} = 7.27$ ppm

Substituent	$\delta_{ortho}—\delta_{C_6H_6}$	$\delta_{meta}—\delta_{C_6H_6}$	$\delta_{para}—\delta_{C_6H_6}$
$—CH_3$	—0.16	—0.09	—0.17
$—C(CH_3)_3$	—0.09	0.01	—0.23
$—C_6H_5$	0.22	0.06	—0.04
N	0.73	0.09	0.02
—C=C—	0.20	—0.04	—0.07
—CH=CH—	0.16	0.00	—0.15
$—NO_2$	0.93	0.26	0.39
—CN	0.25	0.18	0.30
—CHO	0.55	0.19	0.28
$—COCH_3$	0.60	0.11	0.19
$—COC_6H_5$	0.44	0.10	0.19
—COCl	0.81	0.21	0.37
$—CONH_2$	0.69	0.18	0.25
—COOH	0.75	0.14	0.25
$—COOCH_3$	0.71	0.08	0.19
$—COOC_6H_5$	0.88	0.16	0.25
—F	—0.29	—0.02	—0.23
—Cl	0.01	—0.06	—0.12
—Br	0.17	—0.11	—0.06
—I	0.38	—0.23	—0.01
$—NH_2$	—0.08	—0.25	—0.64
—O—	—0.34	—0.04	—0.28
—OH	—0.48	—0.12	—0.48
$—OCH_3$	—0.49	—0.11	—0.44
$—OCOC_6H_5$	—0.11	0.07	—0.10
—SH	—0.03	—0.10	—0.19
$—PO(OCH_3)_2$	0.46	0.14	0.22

Table 44

Coupling constants[1]

|J| in Hz

Structure	J	Notes
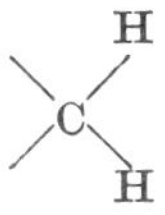 H–C–H (geminal)	10...18	J_{gem} is dependent on angle Φ as well as on other parameters: 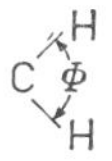 CH_4: $\lvert J \rvert = 12.4$ Hz $\Phi = 109°$; 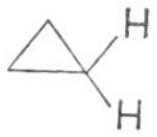 cyclopropane CH_2: $\lvert J \rvert = 4$ Hz $\Phi = 118°$
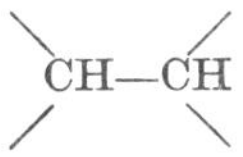 CH—CH	2... 9	J depends on angle Φ [2]: 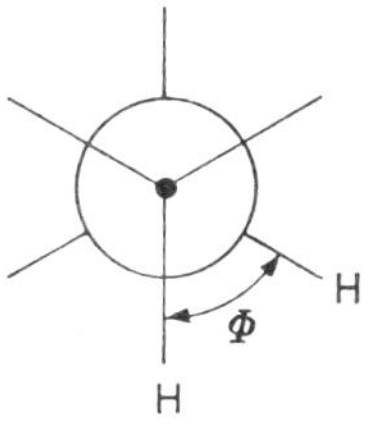 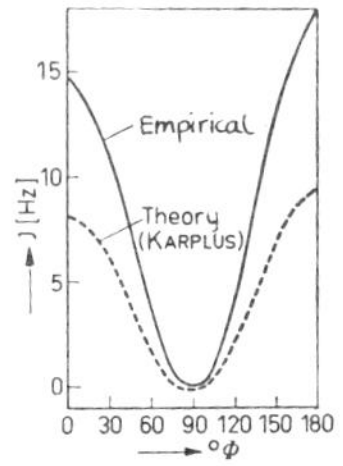
	6... 7	If freely rotatable
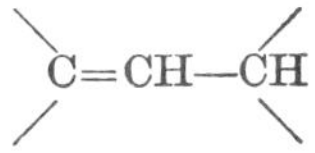 C=CH—CH	4...10	

[1] A. A. Bothner-By in Advances in Magnetic Resonance, ed. J. S. Waugh, Academic Press, New York, London 1965.

[2] M. Karplus: J. Chem. Phys. **30**, 11 (1959).

Table 44 (cont.)

Coupling constants

|J| in Hz

Structure	J	Notes
>C=CH—CH=C<	10...13	
>C=C(H)(H) (geminal)	0... 3.5	J_{gem}, J_{cis} and J_{trans} are very dependent on electronegativity of substituents.
H(>C=C<)H (trans)	12... 18	
H(>C=C<)H (cis)	5... 14	
>CH—CHO	1... 3	
>CH—C≡CH	2... 3	Summary of long-range coupling[3].
HC=C—CH<	0... 3	
HC—C=C—CH	0... 1.6	
cyclohexane H(a), H(e), H(e), H(a)	J_{ee}: 2... 4 J_{ea}: 2... 4 J_{aa}: 6...13	

Table 44 (cont.)

Coupling constants

|J| in Hz

Structure	J	Notes
benzene ring, H–C and H (o, m, p positions)	*o*: 7…10 *m*: 2… 3 *p*: 0… 1	
five-membered ring with X, H_2, H_3, H_4, H_5	0.9…4.7	see table below
benzene ring, F and H (o, m, p positions)	*o*: ∼8 *m*: ∼6 *p*: ∼1	
$\mathrm{CF_2}$ (C with two F)	∼55	
CF—CF	5…20	
P—H	450…550	
P—CH		
P—O—CH	5…15	

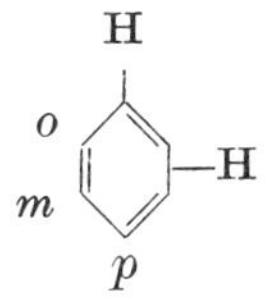

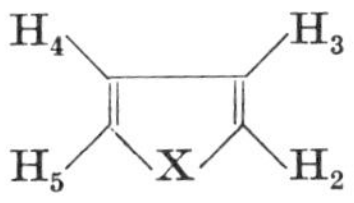

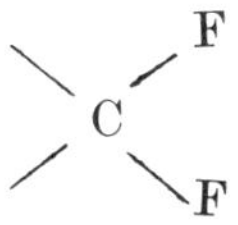

X	J_{23}	J_{34}	J_{24}	J_{25}
O	2.0	3.5	0.9	1.5
N	2.7	3.7	1.3	2.1
S	4.7	3.4	1.0	2.9

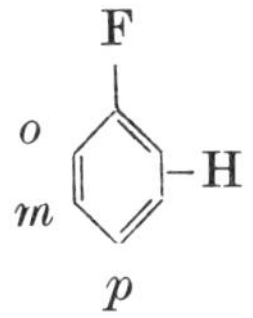

[3] S. STERNHELL: Rev. Pure Appl. Chem. **14**, 15 (1964).

Table 45

Chemical shift of protons in heterocycles

Structure	Compound	Proton	δ [ppm]
	Oxiran	2:	2.7
	Aziridine	1:	0.0
		2:	1.6
	Oxetan	2:	4.7
		3:	2.7
	Tetrahydrofuran	2:	3.7
		3:	1.8
	Pyrrolidine	2:	2.7
		3:	1.6
	Tetrahydrothiophen	2:	2.8
		3:	1.9
	Furan	2:	7.4
		3:	6.4
	Pyrrole	1:	8.0
		2:	6.7
		3:	6.2
	Thiophen	2:	7.3
		3:	7.1
	Pyrazole	1:	12.5
		3:	7.6
		4:	6.3

Table 45 (cont.)

Chemical shift of protons in heterocycles

		Proton	δ [ppm]
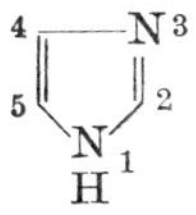	Imidazole	1:	13.4
		2:	7.7
		4:	7.2
		5:	7.2
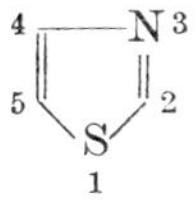	Thiazole	2:	8.8
		4:	8.0
		5:	7.4
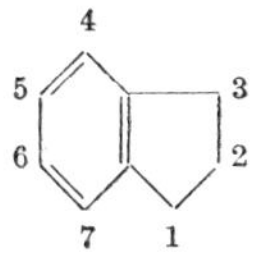	Indane	1:	2.9
		2:	2.0
		3:	2.9
		4...7:	7.2
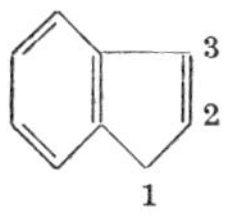	Indene	1:	3.3
		2:	6.5
		3:	6.8
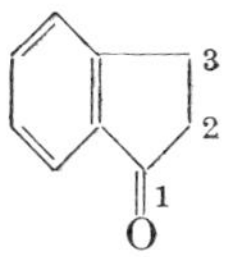	Indan-1-one	2:	2.6
		3:	3.1
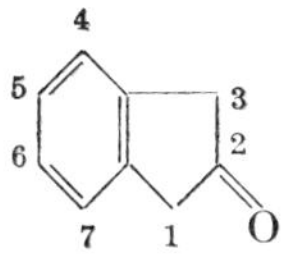	Indan-2-one	1,3:	3.5
		4...7:	7.3
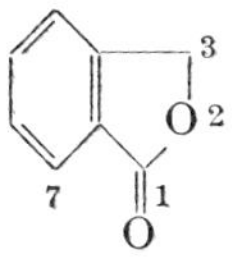	Isocoumaran-1-one	3:	5.3
		7:	7.6
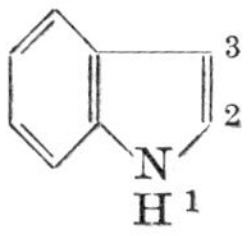	Indole	1:	10.0
		2:	7.3
		3:	6.5

Table 45 (cont.)

Chemical shift of protons in heterocycles

	Proton	δ[ppm]
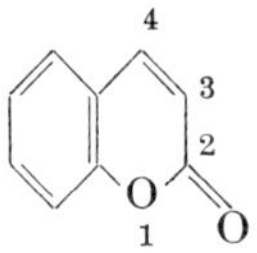 Coumarin	3:	6.5
	4:	7.9
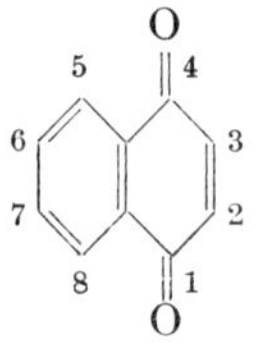 1,4-Naphthoquinone	2.3	7.0
	5.8	8.1
	6.7	7.8
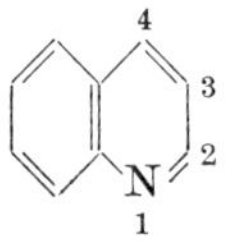 Quinoline	2:	9.0
	3:	7.5
	4:	8.3
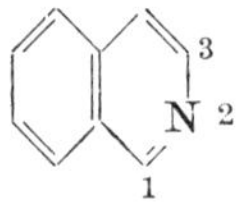 Isoquinoline	1:	9.1
	3:	8.5
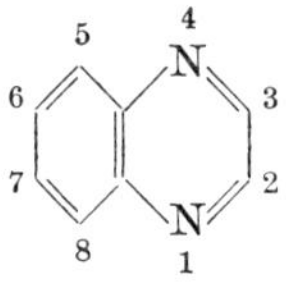 Quinoxaline	2:	8.8
	5:	8.1
	6:	7.8
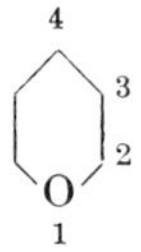 Oxan	2:	3.6
	3:	1.6
	4:	1.6
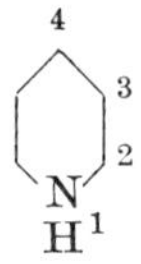 Piperidine	1:	2.0
	2:	2.8
	3:	1.5
	4:	1.5
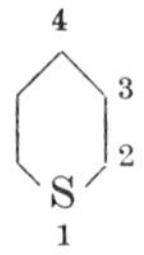 Thian	2:	2.6
	3:	1.7
	4:	1.7

Table 45 (cont.)

Chemical shift of protons in heterocycles

	Proton	δ [ppm]
γ-Pyran	2:	6.2
	3:	4.6
	4:	2.7
Pyridine	2:	8.6
	3:	7.4
	4:	7.8
Morpholine	1:	1.9
	2:	2.9
	3:	3.6
N-Methylpiperazine	1:	2.1
	2:	2.9
	3:	2.4
	7:	2.3
1,2-Dithian	3:	2.7
	4:	1.9
Pyridazine	3:	9.2
	4:	7.5
Pyrimidine	2:	9.3
	4:	8.8
	5:	7.4
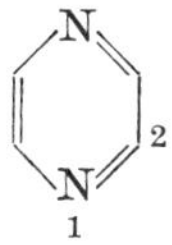 Pyrazine	2:	8.5

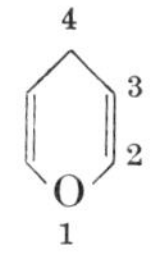

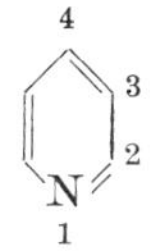

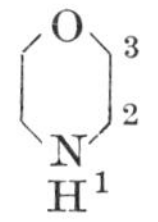

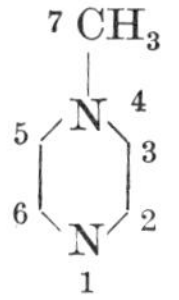

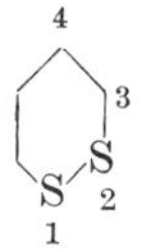

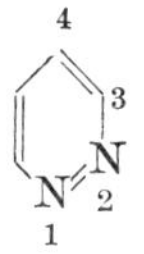

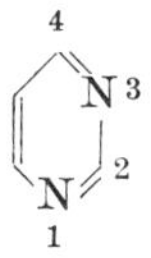

Table 46

Notes on spin-spin interaction in proton resonance

A simple interpretation is possible if the ratio of the difference in chemical shift Δ(Hz) to the coupling constant J (Hz) is evaluated.

Higher order spectra: $$\frac{\Delta}{J} < \sim 10. \qquad (1)$$

Usually these cannot be interpreted without further information. Spectra for various Δ/J ratios are tabulated.*

First order spectra: $$\frac{\Delta}{J} < \sim 10. \qquad (2)$$

The following simple rules hold for a system of type A_mX_n (m magnetically equivalent nuclei A interact with n magnetically equivalent nuclei X), assuming that each of the m nuclei of group A couple in the same way with each of the n nuclei of group X:

a) Nuclei of a group which are magnetically equivalent give no detectable interaction in the spectrum.

b) The multiplicity of the A band is determined by the number n and spin I of the X nuclei according to the formula $2n\mathrm{I} + 1$, so for protons: $n + 1$. If group A interacts at the same time with a third group Y, then the multiplicity of the A band is given by the product of multiplicities caused by X and Y. If the coupling constants J_{AX} and J_{AY} are equal, then the groups X and Y interact with A as if they were equivalent.

c) If A interacts with X, then the multiplet of A forms a symmetrical group of equidistant lines, the intensities of which, when the nuclei X have spin $\frac{1}{2}$, are in the ratio given by the binomial coefficients (e.g. 1:1, 1:2:1, 1:3:3:1, 1:4:6:4:1, 1:5:10:10:5:1, 1:6:15:20:15:6:1 etc.). Because of the sharp decrease in intensity, the outermost lines of a higher multiplet may be invisible.

d) The coupling constant is given by the distance in hertz between two adjacent lines of the multiplet.

e) The chemical shift of a multiplet is given by the position of its midpoint.

f) The coupling constants are independent of the magnetic field strength.

g) Each spin-spin interaction is reciprocal, i.e. the group X band is also split into a multiplet by A, the number of lines being determined by number and spin of A nuclei. Both multiplets have the same coupling constant.

h) If the A and X nuclei are nearly equivalent, the relative intensities of lines within a multiplet no longer exactly correspond to the binomial coefficient (spectra of higher order). The symmetry of the binomial distribution is lost, the lines of A nearer to the X multiplet becoming more intense than those further from X. The same is found with the X multiplet. This effect is useful in the interpretation of spectra.

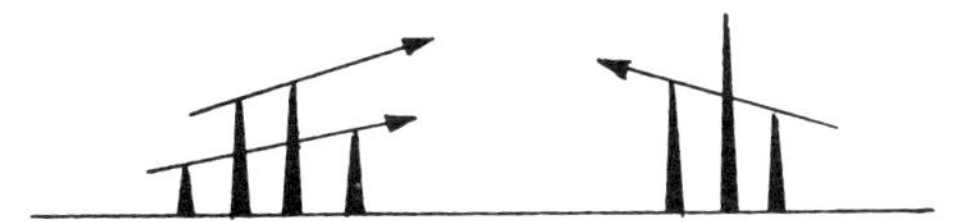

* K. B. Wiberg, B. J. Nist: The Interpretation of NMR Spectra. W. A. Benjamin, Inc. New York 1962.

Table 47

Notes on the proton bands of —OH, —NH and —SH

R—OH		0.5 to 4.5 ppm. At lower field for enols. H-bonded enols usually 11–16 ppm. Bands not always sharp. Their position is dependent on concentration, temperature of solvent and nature of solvent.
	Pure	OH signal should form a multiplet as result of coupling with neighbouring protons, but usually rather diffuse singlets are found. In DMSO, splitting is usually clearly visible.
	Trace of acid	Signal is sharp and because of rapid exchange, lies between frequencies predicted for R—OH and acid protons. Band position is dependent on acid concentration.
HO—C₆H₅ (phenol)		Approx. 4.5 ppm. Intermolecular H-bonding causes a shift of 1 ppm to lower field. If intramolecular H-bonding occurs, chemical shift up to approx. 12 ppm. Position of band is dependent on concentration and temperature.
R—COOH		9.5 to 13 ppm. The position of OH signals in non-polar solvents is independent of concentration (in range 5–10%). Signal position shifted by trace of pyridine.
R—SH		1 to 2 ppm. At lower fields for thioenols (at approx. 5 ppm). In this case the bands may be broad. The band position is dependent on concentration and temperature of solution and on nature of solvent.

$R—NH_2$ $R—NH—R'$	1 to 4 ppm. Lines mostly broad. (benzene ring)—NH_2: 3.5 to 6 ppm.	
	Pure	Bands not split because of exchange. Signal usually sharp but sometimes broad. Position dependent on solvent.
	Trace of acid	Signal lies between the position predicted for —NH— and acid proton.
$R—NH_2$ + conc. acid $R—NH^+_3$	Complete protonation inhibits exchange and the signal of the proton bonded to nitrogen is split into three bands of equal intensity as result of coupling with nitrogen (I = 1) (large coupling constant: J_{NH} = approx. 50 Hz).	
$R\ R'—NH^+_2$ $R_3—NH^+$	Sharp band at low field. The nitrogen relaxation is large for protonated secondary and tertiary amines, so no coupling with linked protons can be detected.	
$R—CO—NH_2$	5 to 8.5 ppm. Bands are often very broad (even undetectable). Sharper signals can be obtained by alkaline catalysis of proton exchange. The position of signals is dependent on solvent.	
$R—CO—NH—R'$	Signals as above. Usually sharper than signals of primary amides. The splitting of the protons vicinal to the NH group is often clearly visible, even if the signal of the NH group is broad and unstructured.	
$R—CO—NH—CO—R'$	Approx. 9 to 12 ppm. Broad bands.	
$=N—OH$	Approx. 10 to 12 ppm. Bands may be broad.	

Table 48

Proton resonance spectroscopy

Selection of literature

Textbooks and monographs

BHACCA, N. S., and D. H. WILLIAMS: Applications of NMR Spectroscopy in Organic Chemistry. Holden-Day, Inc., San Francisco, London, Amsterdam 1964.

BIBLE, R. H., jr.: Interpretation of NMR Spectra. Plenum Press, New York 1965.

EMSLEY, J. W., J. FEENEY and L. H. SUTCLIFFE: High Resolution Nuclear Magnetic Resonance Spectroscopy, Volume 1 & 2. Pergamon Press, Oxford, London, Edinburgh, New York, Paris, Frankfurt 1966.

FLUCK, E.: Die kernmagnetische Resonanz und ihre Anwendung in der anorganischen Chemie. Springer-Verlag, Berlin, Göttingen, Heidelberg 1963.

JACKMAN, L. M.: Applications of Nuclear Magnetic Resonance Spectroscopy in Organic Chemistry. Pergamon Press, London, Oxford, New York, Paris 1959.

*BIBLE, R. H.: Guide to the NMR Empirical Method. Plenum Press, New York 1965.

POPLE, J. A., W. G. SCHNEIDER and H. J. BERNSTEIN: High Resolution Nuclear Magnetic Resonance. McGraw-Hill Book Company, Inc., New York, Toronto, London 1959.

ROBERTS, J. D.: An Introduction to the Analysis of Spin-Spin Splitting in High-Resolution Nuclear Magnetic Resonance Spectra. W. A. Benjamin, Inc., New York 1961.

STREHLOW, H.: Magnetische Kernresonanz und chemische Struktur. Dr. Dietrich Steinkopff Verlag, Darmstadt 1962.

*SUHR, H.: Anwendungen der Kernmagnetischen Resonanz in der organischen Chemie. Springer-Verlag, Berlin, Heidelberg, New York 1965.

* Especially to be recommended for the interpretation of spectra of organic compounds.

Proton resonance spectroscopy

Selection of literature

Collection of data

HERSHENSON, H. M.: Nuclear Magnetic Resonance and Electron Spin Resonance Spectra, Index for 1958–1963. Academic Press, New York, London 1965.

Nuclear Magnetic Resonance Spectral Data. American Petroleum Institute, Research Project 44, Texas A & M University, College Station, Texas.

BHACCA, N. S., D. P. HOLLIS, L. F. JOHNSON, and E. A. PIER: NMR Spectra Catalog, Volume I, II. Varian Associates, Palo Alto, California 1962, 1963.

BOVEY, F. A.: NMR Data Tables for Organic Compounds, Volume I. Interscience/Wiley, New York, London, Sydney 1968.

BRÜGEL, W.: Kernresonanz-Spektrum und chemische Konstitution, Band I: Die spektralen Kernresonanzparameter von Verbindungen mit analysiertem Spektrum. Verlag Steinkopff, Darmstadt 1968.

HOWELL, M. G., A. S. KENDE, and J. S. WEBB: Formula Index to NMR Literature Data, Volume 1, 2. Plenum Press, New York 1965, 1966.

NEUDERT, W., and H. RÖPKE: Steroid-Spektrenatlas. Springer Verlag, Berlin, Heidelberg, New York, London, Sydney 1968.

NMR Spectra Catalog. Sadtler Research Laboratories, Philadelphia, Pennsylvania.

SZYMANSKY, H. A., and R. E. YELIN: NMR Band Handbook. IFI/Plenum, New York, Washington 1968.

TIERS, G. V. D.: Table of τ-values for a Variety of Organic Compounds. Central Research Department Minnesota Mining and Manufacturing Company, St Paul, Minnesota.

Table 50

Mass Correlations

m/e	Fragment x	M^+—x
12	C	
13	CH	
14	CH_2, N, N_2, CO	
15	CH_3	M^+—15 non-specific, CH_3 at higher intensity
16	O, NH_2, O_2, NH_4	M^+—16 rarely CH_4, (but R^+—CH_4 common in alkyl fragments)
17	OH, NH_3	M^+—17 non-specific O indication, NH_3 from prim. amines
18	H_2O, NH_4	M^+—18 non-specific O indication, strong for many alcohols, some acids, ethers and lactones
19	H_3O, F	M^+—19 F indication
20	HF, Ar^{++}	M^+—20 F indication
21	$C_2H_2O^{++}$ (rare)	
22	CO_2^{++}	
23	Na (rare)	
24	C_2	
25	C_2H	M^+—25 rare for terminal —C≡CH group
26	C_2H_2, CN	M^+—26 from pure arom. compounds, rare from cyanides
27	C_2H_3, HCN	M^+—27 CN from cyanides, C_2H_3 from terminal vinyl groups and some ethyl esters
28	C_2H_4, N_2, CO	M^+—28 CO from arom. linked O, ethylene from cyclohexenes by retro-Diels-Alder reaction, from alkyl residues by H-rearrangement, non-specifically from alicyclic compounds
29	C_2H_5, CHO	M^+—29 arom. linked O, non-specific for hydrocarbons
30	C_2H_6, HN—CH_3, CH_2O, (N fragment), BF	M^+—30 CH_2O from cycl. ethers and arom. methyl ethers, NO from nitro-compounds and nitro-esters
31	CH_3O, CH_2OH, CH_3NH_2, CF, (O fragment)	M^+—31 methyl ethers, methyl esters, alcohols
32	O_2, CH_3OH, S	M^+—32 methyl esters, some sulphides and methyl ethers
33	CH_3OH_2, SH	M^+—33 SH non-specific S indication, (M^+—18—15) non-specific O indication, strong for alcohols
34	SH_2, (S fragment)	M^+—34 non-specific S indication, strong for thiols

Table 50 (cont.)

Mass Correlations

m/e	Fragment x	M^+—x
35	^{35}Cl, SH_3	M^+—35 chlorides, nitrophenyl compounds (M^+—17—18)
36	HCl, C_3	M^+—36 chlorides
37	^{37}Cl, C_3H_2	
38	$H^{37}Cl$, C_3H_2	
39	C_3H_3	M^+—39 weak for arom. hydrocarbons
40	Ar, C_3H_4	M^+—40 rare for CH_2CN
41	C_3H_5, CH_3CN	M^+—41 C_3H_5 from alicycl, compounds, CH_3CN from arom. *N*-methyl and *o*—C methyl heterocycles
42	$CH_2{=}C{=}O$, C_3H_6, C_2H_4N	M^+—42 non-specific for aliph. and alicycl. systems, strong from cyclohexenes from retro-Diels-Alder reaction, by arrangement from $\alpha\beta$-cyclohexones, enol- and enamine acetates
43	CH_3CO, C_3H_7, C_2H_4N, CONH	M^+—43 acetyl, propyl, arom. methyl ethers (M^+—15—28), non-specific for aliph. and alicycl. systems
44	CO_2, CH_3NHCH_2, CH_2CHOH, (N fragments)	M^+—44 CO_2 from acids, esters, butane from aliph. hydrocarbons
45	C_2H_5O, HCS, (sulphides)	M^+—45 ethyl esters, ethyl ethers, lactones, CO_2H from some esters
46	C_2H_5O, NO_2	M^+—46 ethyl esters, ethyl ethers, rarely acids, nitro compounds, n-alkanols
47	CH_3S, $C^{35}Cl$, $C_2H_5OH_2$	
48	CH_3SH, $CH^{35}Cl$	
49	$C^{37}Cl$, $CH_2{}^{35}Cl$	
50	C_4H_2, $CH_3{}^{35}Cl$	
51	C_4H_3 } aromatic fragments	
52	C_4H_4, $CH_3{}^{35}Cl$ } aromatic fragments	
53	C_4H_5 } aromatic fragments	
54	//‾\\, C_2H_4CN	M^+—54 cyclohexene (retro-Diels-Alder reaction)
55	C_4H_7, C_2H_3CO	M^+—55 C_4H_7 from alicycl. systems and butyl esters
56	C_4H_8, C_3H_4O	M^+—56 } non-specific for alkanes and alkanes and alicycl. systems
57	C_4H_9, C_2H_5CO	M^+—57 } non-specific for alkanes and alkanes and alicycl. systems
58	CH_3COHCH_2, C_2H_5—$CHNH_2$, $C_2H_6NCH_2$	M^+—58 C_3H_6O from α-methylaldehydes
59	C_2H_6COH, $C_2H_5OCH_2$, CO_2CH_3, CH_3CONH_2	M^+—59 methyl esters

Table 50 (cont.)

Mass Correlations

m/e	Fragment x	M^+—x
60	$CH_2CO_2H_2$, CH_2ONO	M^+—60 θ-Acetate (M^+—Ac.OH), methyl esters (M^+—CH_3OH —CO)
62	$HOCH_2CH_2OH$	M^+—62 ethylene ketals
63	C_5H_3	
65	C_5H_5	
66		
67	 O	
68	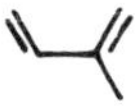 , C_4H_4O, C_3H_6CN	
69	C_5H_9, C_3H_5CO, CF_3, C_3HO_2 (1,3-dioxyaromatics)	
70	C_5H_{10}  N H	
71	C_5H_{11}, C_4H_7CO	
72	$C_4H_{10}N$, $C_3H_7NHCH_2$, $C_2H_5COHCH_2$	
73	$CO_2C_2H_5$, $C_3H_7OCH_2$, $CH_2CO_2CH_3$, C_4H_8OH (O fragment)	
74	$CH_2COHOCH_3$	
75	$C_2H_5CO_2H_2$, $C_2H_5SCH_2$, $CH_3OCHOCH_3$ (dimethyl-acetate)	
76	C_6H_4	
77	C_6H_5	
78	C_6H_6	
79	C_6H_7, ^{79}Br	
80	C_6H_8, $H^{79}Br$ CH_3S_2H   N H, N H CH_2,	
81	C_6H_9, ^{81}Br 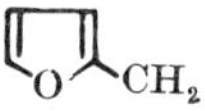 O CH_2	
82	C_6H_{10}, $H^{81}Br$,  N H	
83	C_6H_{11}, C_4H_9CO  N CH_3	
85	C_6H_{13}, C_4H_9CO	

Table 50 (cont.)

Mass Correlations

m/e	Fragment x	M^+—x
86	$C_3H_7COH=CH_2$	
87	$CO_2C_3H_7$, $CH_2CO_2C_2H_5$, $CH_2CH_2CO_2CH_3$ (O fragment)	
88	$CH_2=COHOC_2H_5$, $CH_3CH=COHOCH_3$	
91	  CH_2 Cl n-alkyl chlorides	
92	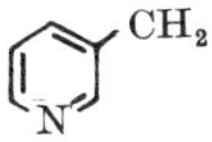 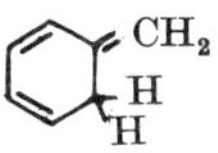 CH_2 N CH_2 H H	
93	$CH_2{}^{79}Br$, 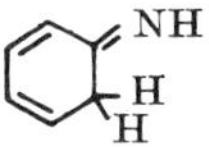 NH H H	
94	$CH_3{}^{79}Br$, 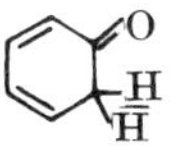 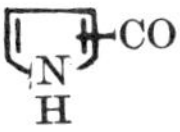 O H H N H CO	
95	$CH_2{}^{81}Br$ 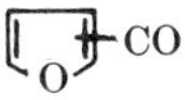 O CO	
96	$C_5H_{10}CN$, $CH_3{}^{81}Br$	
97	C_7H_{13},  S CH_2	
98	 N CH_3	
99	C_7H_{15}, 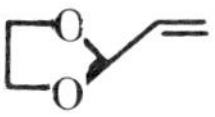 O O ethylene ketals	
104	C_2H_5CHONO 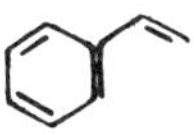 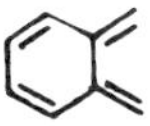	
105	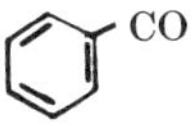 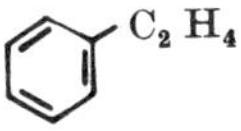 CO C_2H_4	
111	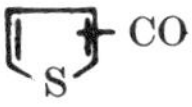 S CO	
115	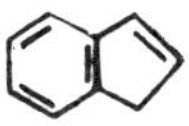	

Table 50 (cont.)

Mass Correlations

m/e	Fragment x	M^{+}—x
119	C_2F_5 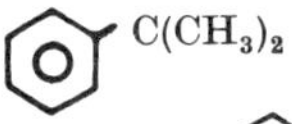 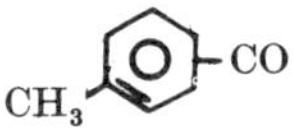 $C(CH_3)_2$ CH_3 CO	
120	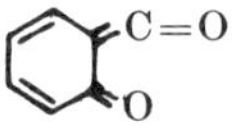 C=O O	
121	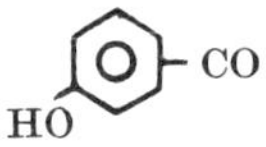 HO CO	
127	I, 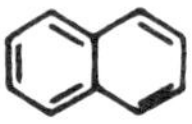	
128	HI,	
130	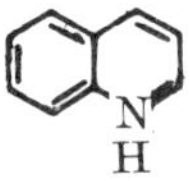 N H	
131	C_3F_5	
135	 Br (n-alkyl bromides)	
141	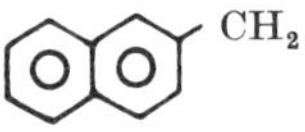 CH_2	
144	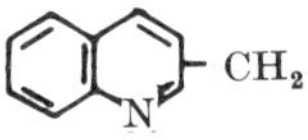 N CH_2	
149	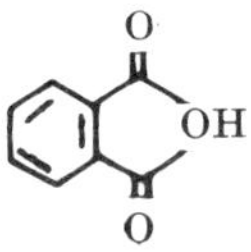 O OH (phthalates) O	
152		

Table 51

Natural abundance of isotopes

(relative to abundance of isotope with lowest mass)

Element	Mass P of isotope of lowest mass	Isotope mass $P+1$ Abundance (%)	Isotope mass $P+2$ Abundance (%)
C	12	1.11	—
H	1	0.015	—
O	16	—	0.2
N	14	0.37	—
S	32	0.78	4.4
F	19	—	—
Cl	35	—	32.5
Br	79	—	97.9

The relative intensity of the isotope peaks, caused by the presence of one atom of type A (mass of lightest isotope: P_A, abundance of isotope $P_A + 1{:}x$, of mass $P_A + 2{:}y$) and of one atom of type B (mass of lightest isotope P_B, abundance of isotope $P_B + 1{:}r$, of mass $P_B + 2{:}s$), can be worked out from the scheme on page 88.

Table 51 (cont.)

Mass of isotope		P_A	$P_A + 1$	$P_A + 2$		
	Rel. abundance of isotopes of B: ↓ of isotopes of A: →	1	x	y		
P_B	1	**1 · 1**	**1 · x**	**1 · y**		
$P_B + 1$	r		**r · 1**	**r · x**	**r · y**	
$P_B + 2$	s			**s · 1**	**s · x**	**s · y**
Sum = rel. intensity of isotope peaks of A + B		1	$x + r$	$y + (r \cdot x) + s$	$(r \cdot y) + (s \cdot x)$	$s \cdot y$
Mass relative to mass M of species containing only light isotopes		M	$M + 1$	$M + 2$	$M + 3$	$M + 4$

The rapidly increasing quantities are formed by successive multiplication of the isotope abundance row by the members of the isotope abundance column, with corresponding shift to the right.
In order to extend the scheme to combinations of 3 or more atoms, the relative intensity of isotope peaks of any combination of atoms can be treated like the relative abundance of a single atom.

Table 52

Isotope peaks of fragments containing chlorine and bromine

(The mass of the species containing only light isotopes is represented by M)

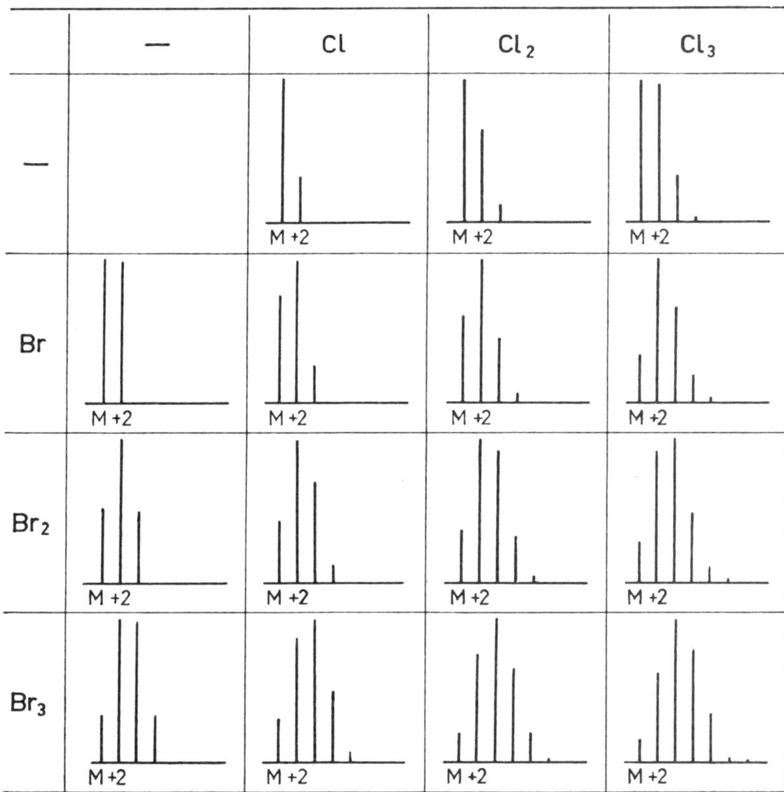

Table 53

Mass spectroscopy

Selection of literature

Textbooks and monographs

BENZ, W.: Massenspektrometrie Organischer Verbindungen. Akademische Verlagsgesellschaft, Frankfurt/Main, 1969.

BEYNON, J. H.: Mass Spectrometry and its Applications to Organic Chemistry. Elsevier Publishing Company, Amsterdam, London, New York 1960.

BEYNON, J. H. and A. E. WILLIAMS: Mass and Abundance Tables for Use in Mass Spectrometry. Elsevier Publishing Company, Amsterdam, London, New York 1963.

BEYNON, J. H., R. A. SAUNDERS and A. E. WILLIAMS: The Mass Spectra of Organic Molecules. Elsevier Publishing Company, Amsterdam, London, New York 1968.

BIEMANN, K.: Mass Spectrometry Organic Chemical Applications. McGraw-Hill Book Company, Inc., New York, San Francisco, Toronto, London 1962.

BRUNNÉE, C., and H. Voshage: Massenspektrometrie. Verlag Karl Thiemig KG., München 1964.

*BUDZIKIEWICZ, H., C. DJERASSI and D. H. WILLIAMS: Mass Spectrometry of Organic Compounds. Holden-Day, Inc., San Francisco, Cambridge, London, Amsterdam 1967.

BUDZIKIEWICZ, H., C. DJERASSI and D. H. WILLIAMS: Structure Elucidation of Natural Products by Mass Spectrometry, Vol. I: Alkaloids; Vol. II: Steroids, Terpenoids, Sugars and Miscellaneous Classes. Holden-Day, Inc., San Francisco 1964.

FIELD, F. H. and J. L. FRANKLIN: Electron Impact Phenomena and the Properties of Gaseous Ions. Academic Press, Inc., Publishers, New York 1957.

HILL, H. C.: Introduction to Mass Spectrometry. Heyden and Sons Ltd., London 1966.

KIENITZ, H.: Massenspektrometrie. Verlag Chemie GmbH., Weinheim 1968.

McLAFFERTY, F. W.: Mass Spectrometry of Organic Ions. Academic Press, Inc., Publishers, New York 1963.

McLAFFERTY, F. W.: Mass Spectral Correlations, (Advances in Chemistry Series 40). American Chemical Society, Washington 1963.

---

* Especially to be recommended for the interpretation of spectra of organic compounds.

Table 54

MS

Mass spectroscopy

Selection of literature

Collection of data

ASTM Index of Mass Spectral Data, Heyden & Sons Limited, London.

Beynon, J. H. and A. E. Williams: Mass and Abundance Tables for Use in Mass Spectrometry. Elsevier Publishing Company, Amsterdam, London, New York 1963.

Catalog of Mass Spectral Data, Manufacturing Chemists Association. Research Project, Carnegie Institute of Technology, Pittsburgh, Pa.

Catalog of Mass Spectral Data, American Petroleum Institute. Research Project 44, Carnegie Institute of Technology, Pittsburgh, Pa.

Compilation of Mass Spectral Data, A. Cornu & R. Massot, Heyden & Sons Limited, London.

Mass Spectral Data Sheets, Mass Spectrometry Data Centre, AWRE, Aldermaston, Berks, England.

Examples

General notes on spectra

IR: All technical details of measurement are given with each spectrum.

NMR: The instrument used, solvent, sweep width and if necessary sweep offset are given with the spectra. TMS was used as a reference throughout, except in Problem 16 in which DMS was used. The arrow in the top right corner of the spectrum gives the direction of increase of magnetic field Hz in z direction. The CPS scale is accordingly in the opposite direction. The instrumental data necessary for interpretation of the spectra are:

Instrument	Resonance frequency for protons	ppm scale related to sweep width of	1 ppm corresponds to
VARIAN A-60	60 MHz	500 Hz	60 Hz
VARIAN HA 100	100 MHz	1000 Hz	100 Hz

MS: The mass/charge ratio is given on the abscissa and the relative intensity as percentage of intensity of strongest signal on the ordinate. 'Metastable peaks' are indicated by an arrow on the m/e scale. A 'metastable peak' appears at mass $m^* = (m_2)^2/m_1$ for the metastable transition $m_1 \rightarrow m_2 + \mathrm{R}$.

UV: All necessary technical details of measurement are given.

The examples are arranged randomly, not in order of increasing difficulty. In certain examples it is not possible to deduce an unequivocal structure from the data given.

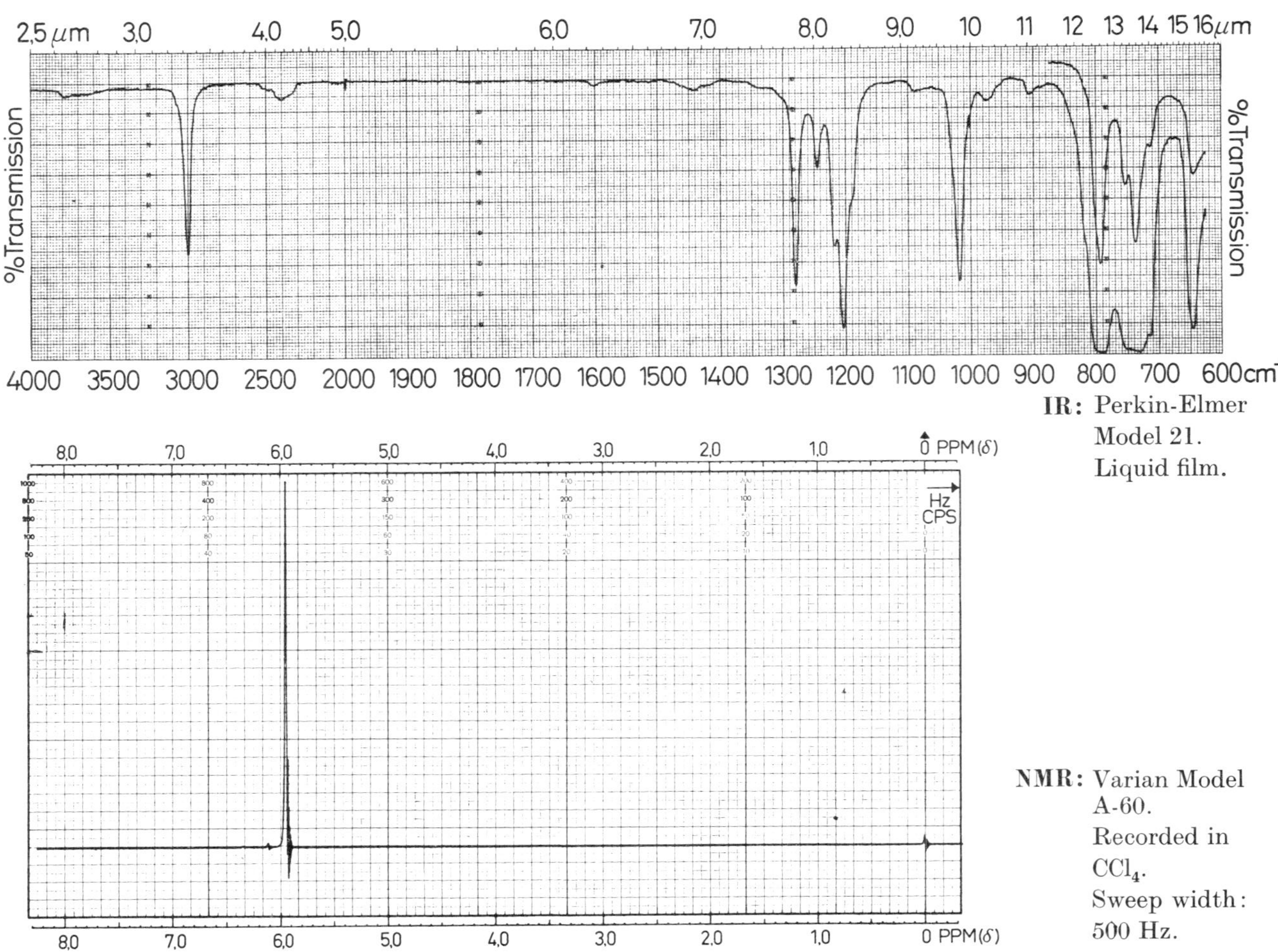

IR: Perkin-Elmer Model 21. Liquid film.

NMR: Varian Model A-60. Recorded in CCl_4. Sweep width: 500 Hz.

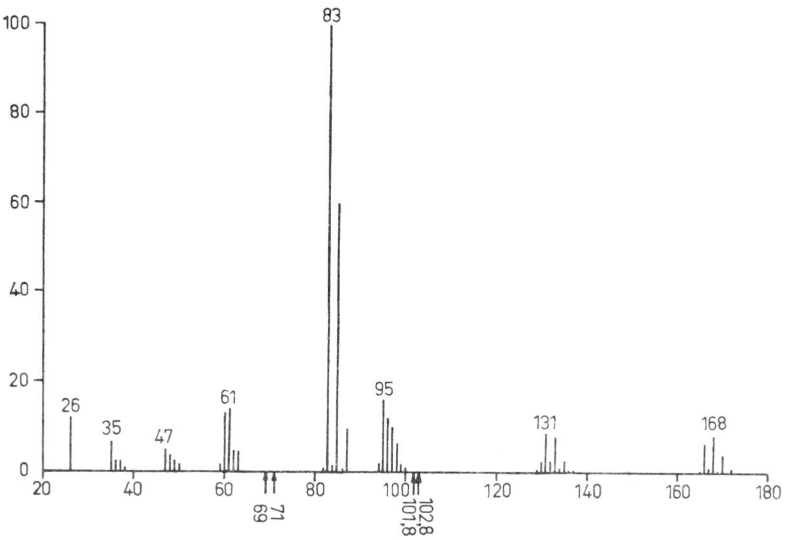

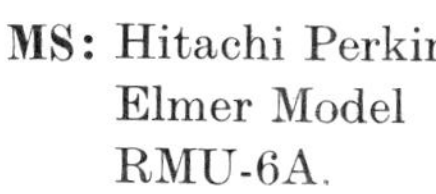

MS: Hitachi Perkin-Elmer Model RMU-6A.

UV: Perkin-Elmer Model 137 UV. Recorded in C_2H_5OH. >210 nm end absorption.

Problem 1

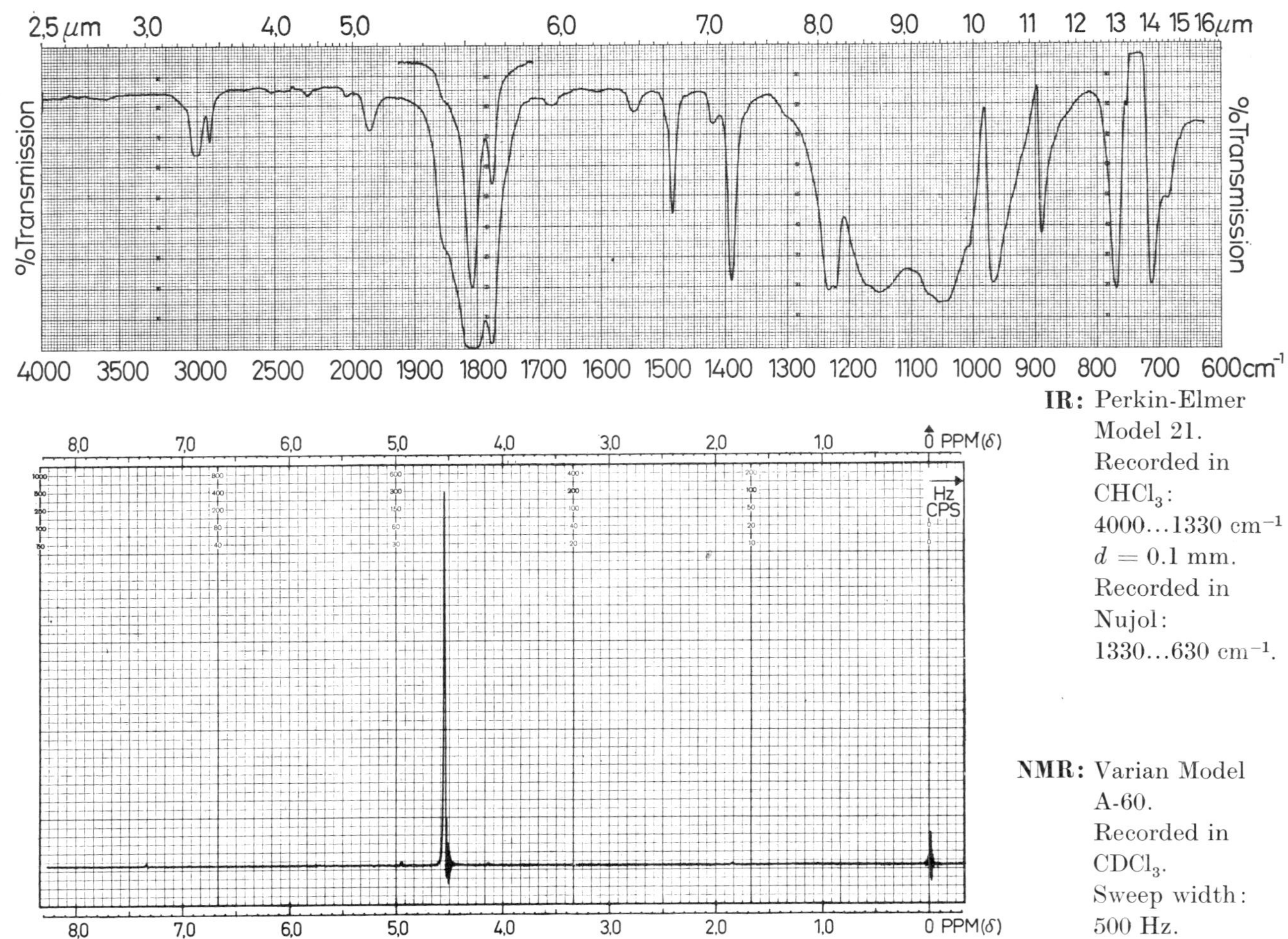

IR: Perkin-Elmer Model 21. Recorded in $CHCl_3$: 4000…1330 cm^{-1} $d = 0.1$ mm. Recorded in Nujol: 1330…630 cm^{-1}.

NMR: Varian Model A-60. Recorded in $CDCl_3$. Sweep width: 500 Hz.

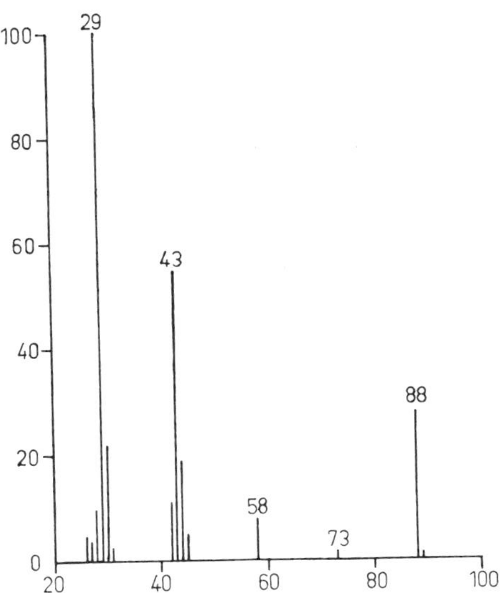

MS: Hitachi Perkin-Elmer Model RMU-6D.

UV: Perkin-Elmer Model 137 UV. Recorded in C_2H_5OH. >210 nm none

Problem 2

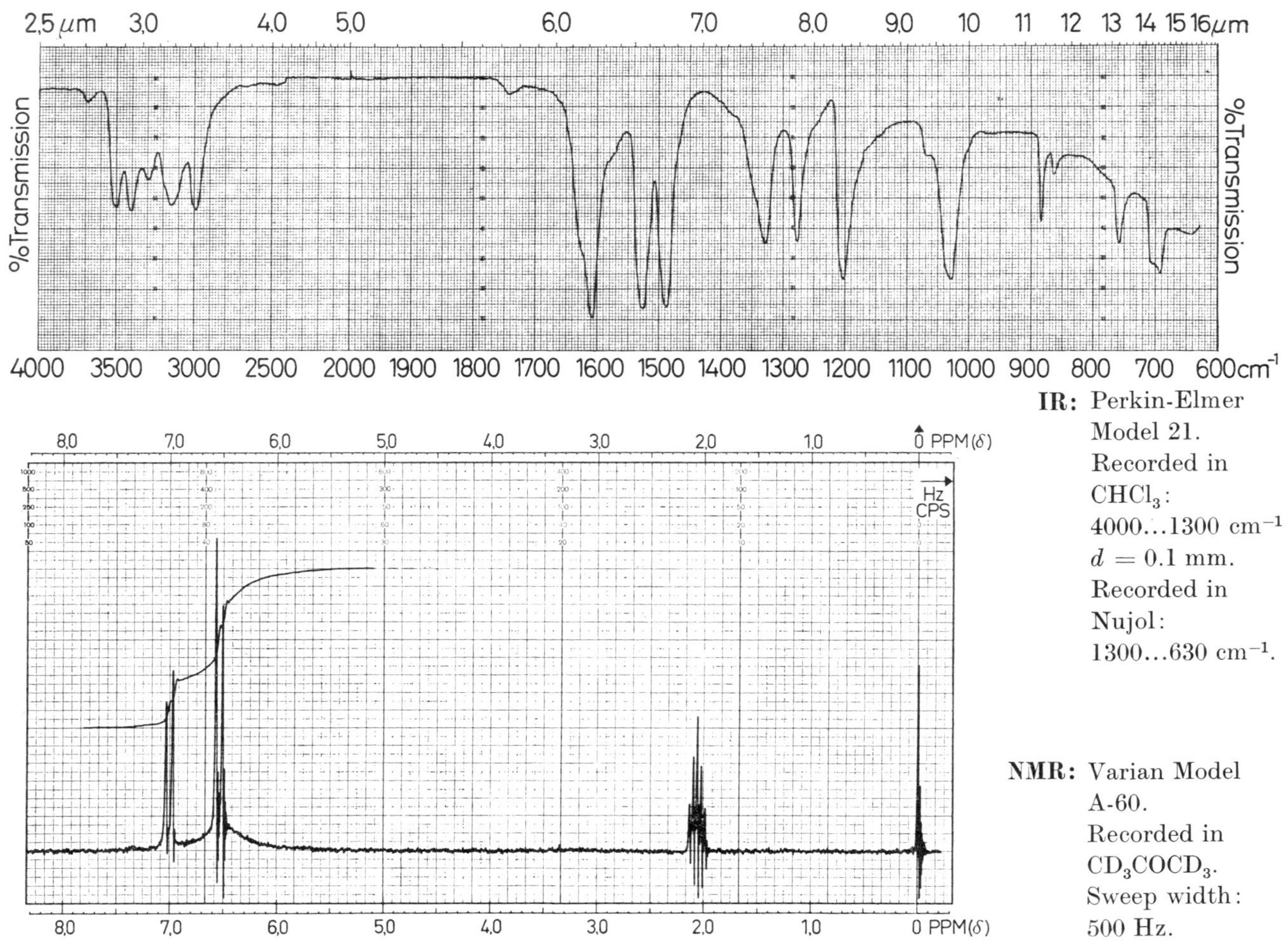

IR: Perkin-Elmer Model 21. Recorded in $CHCl_3$: 4000…1300 cm^{-1} $d = 0.1$ mm. Recorded in Nujol: 1300…630 cm^{-1}.

NMR: Varian Model A-60. Recorded in CD_3COCD_3. Sweep width: 500 Hz.

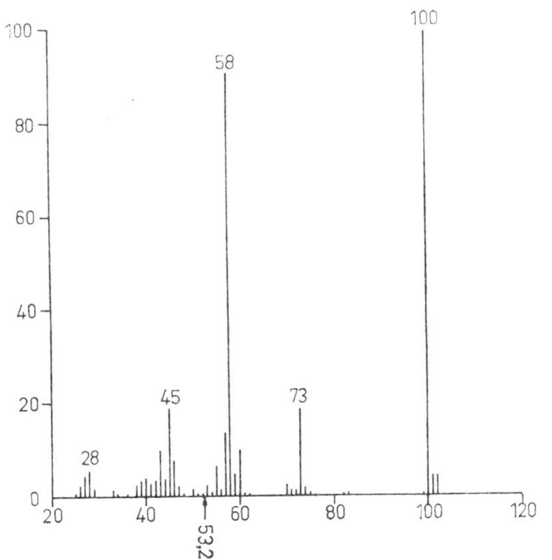

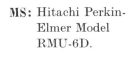

MS: Hitachi Perkin-Elmer Model RMU-6D.

UV: Perkin-Elmer Model 137 UV. Recorded in C_2H_5OH.

λ_{max}	$\log \varepsilon$
256 nm	3.8

Problem 3

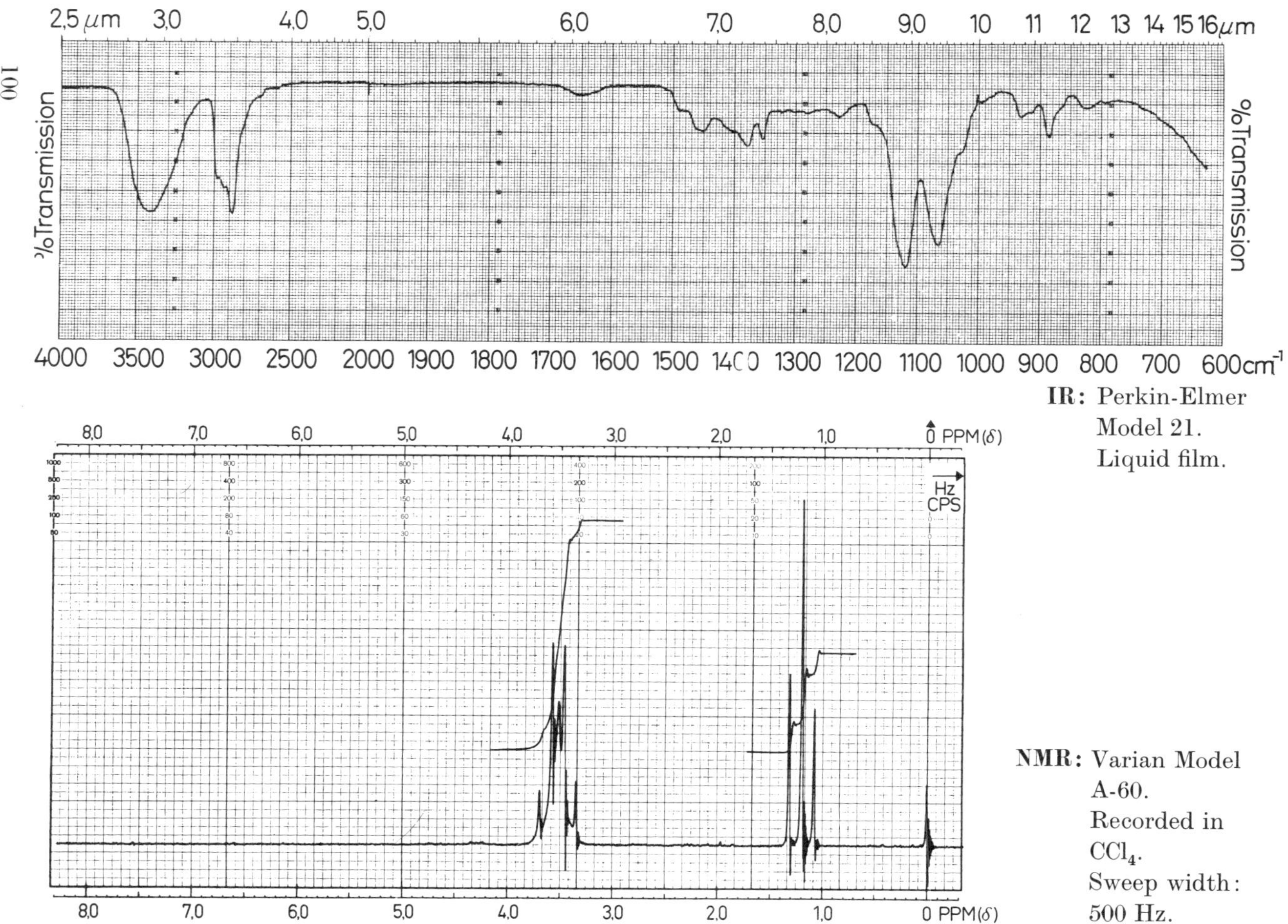

IR: Perkin-Elmer Model 21. Liquid film.

NMR: Varian Model A-60. Recorded in CCl_4. Sweep width: 500 Hz.

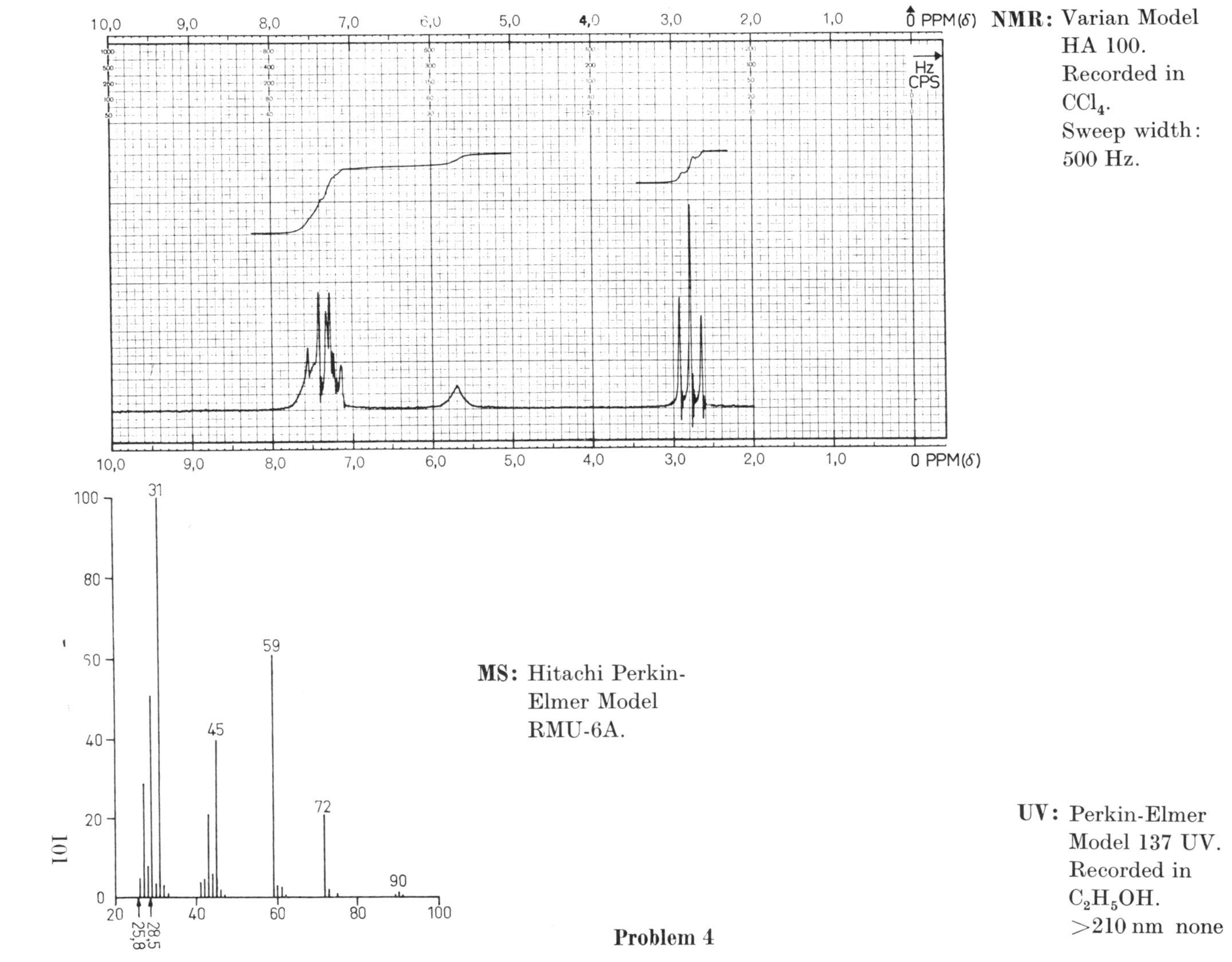

NMR: Varian Model HA 100. Recorded in CCl_4. Sweep width: 500 Hz.

MS: Hitachi Perkin-Elmer Model RMU-6A.

UV: Perkin-Elmer Model 137 UV. Recorded in C_2H_5OH. >210 nm none

Problem 4

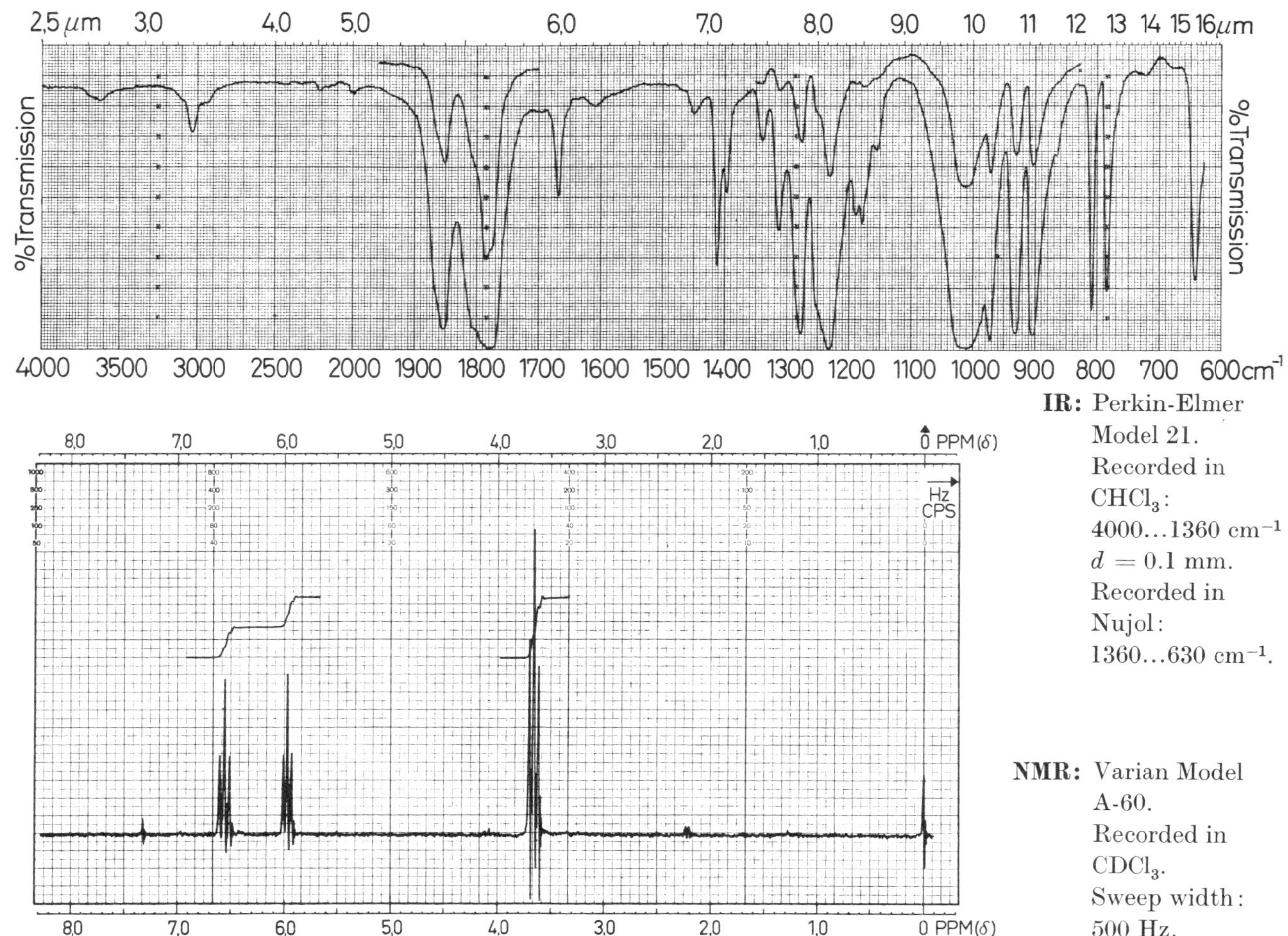

IR: Perkin-Elmer Model 21. Recorded in $CHCl_3$: 4000...1360 cm^{-1} $d = 0.1$ mm. Recorded in Nujol: 1360...630 cm^{-1}.

NMR: Varian Model A-60. Recorded in $CDCl_3$. Sweep width: 500 Hz.

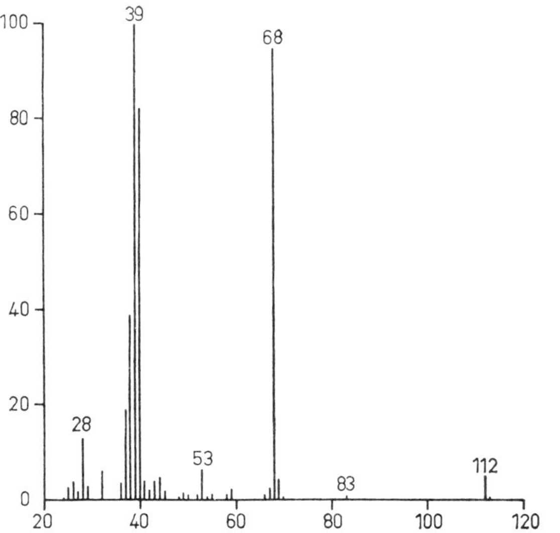

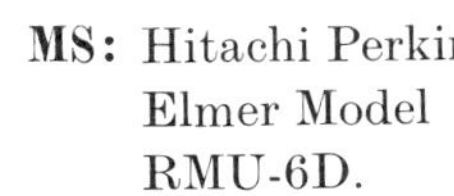

MS: Hitachi Perkin-Elmer Model RMU-6D.

UV: Perkin-Elmer Model 137 UV. Recorded in C_2H_5OH. >210 nm end absorption.

Problem 5

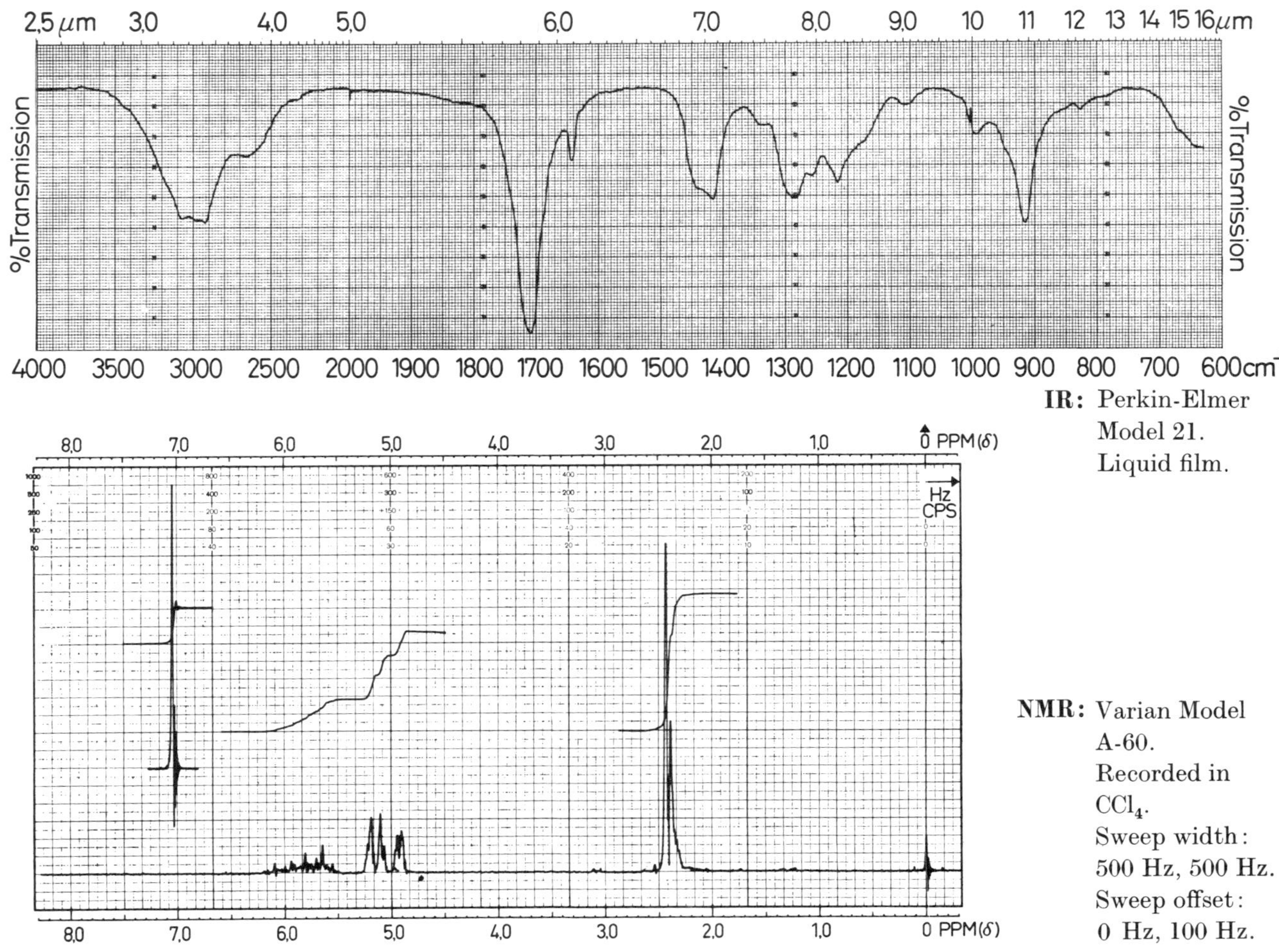

IR: Perkin-Elmer Model 21. Liquid film.

NMR: Varian Model A-60. Recorded in CCl_4. Sweep width: 500 Hz, 500 Hz. Sweep offset: 0 Hz, 100 Hz.

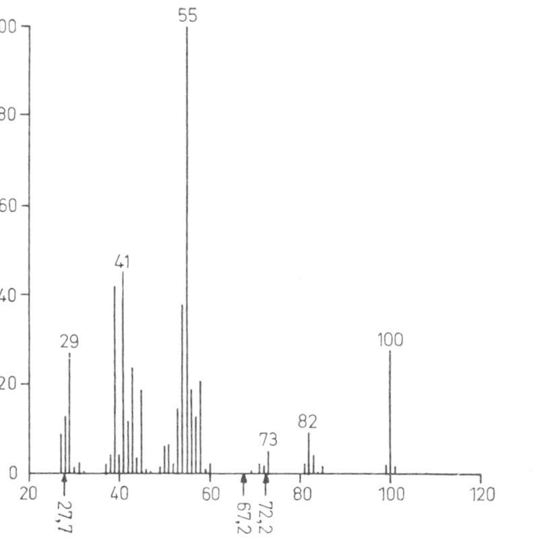

MS: Hitachi Perkin-Elmer Model RMU-6D.

UV: Perkin-Elmer Model 137 UV. Recorded in C_2H_5OH. >210 nm end absorption.

Problem 6

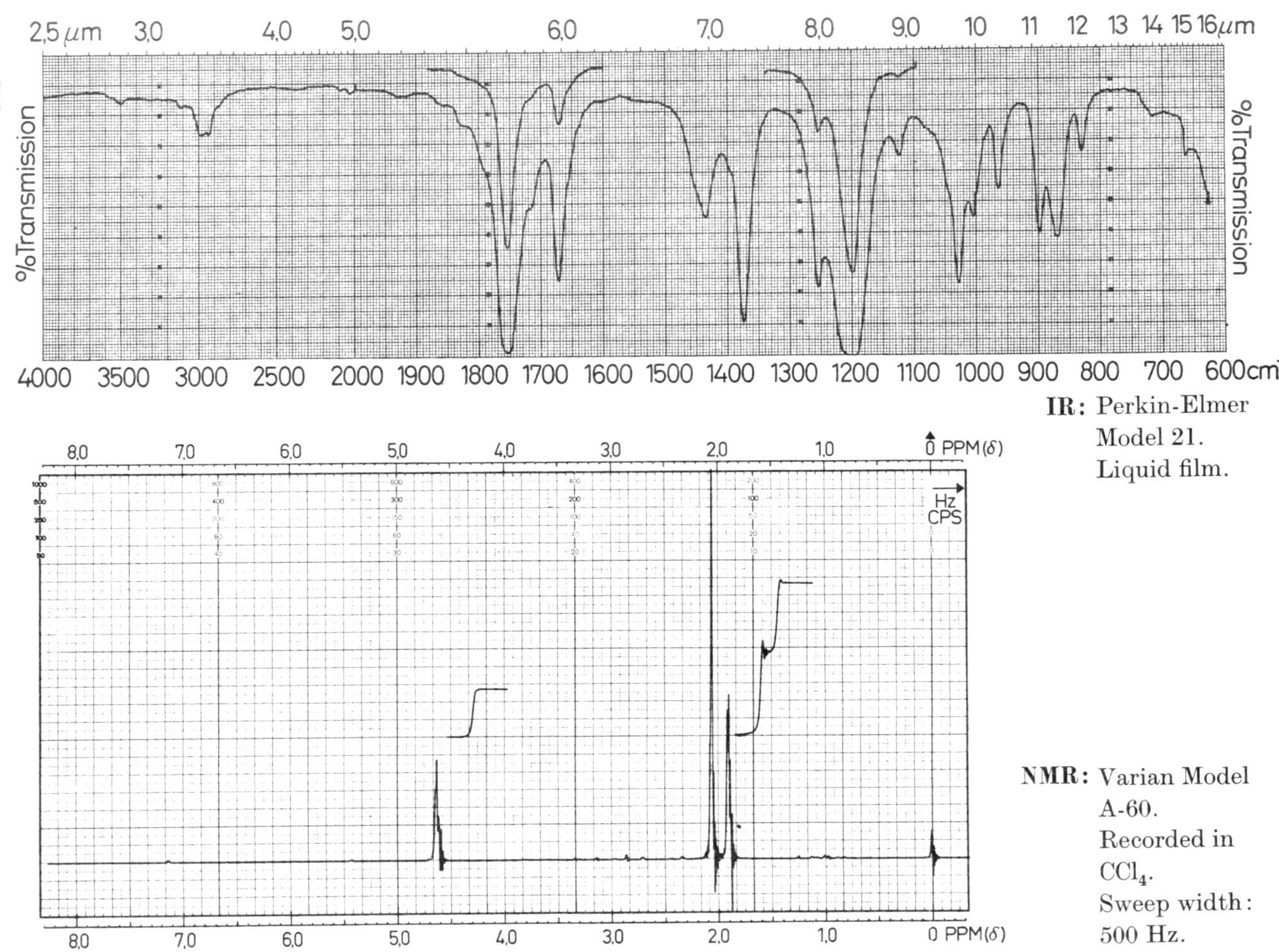

IR: Perkin-Elmer
Model 21.
Liquid film.

NMR: Varian Model
A-60.
Recorded in
CCl_4.
Sweep width:
500 Hz.

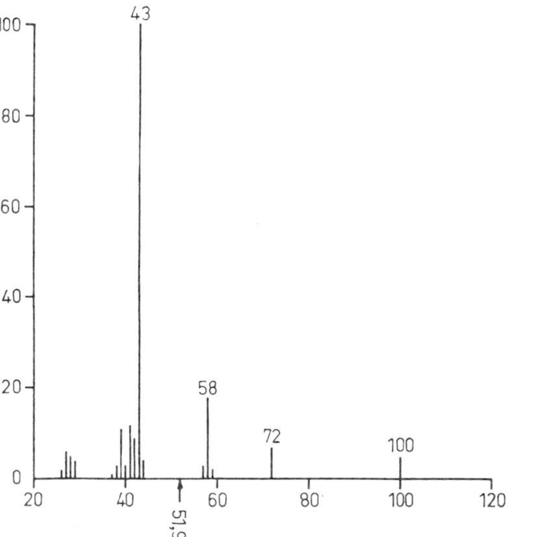

MS: Hitachi Perkin-Elmer Model RMU-6A.

UV: Perkin-Elmer Model 137 UV. Recorded in C_2H_5OH. >210 nm end absorption.

Problem 7

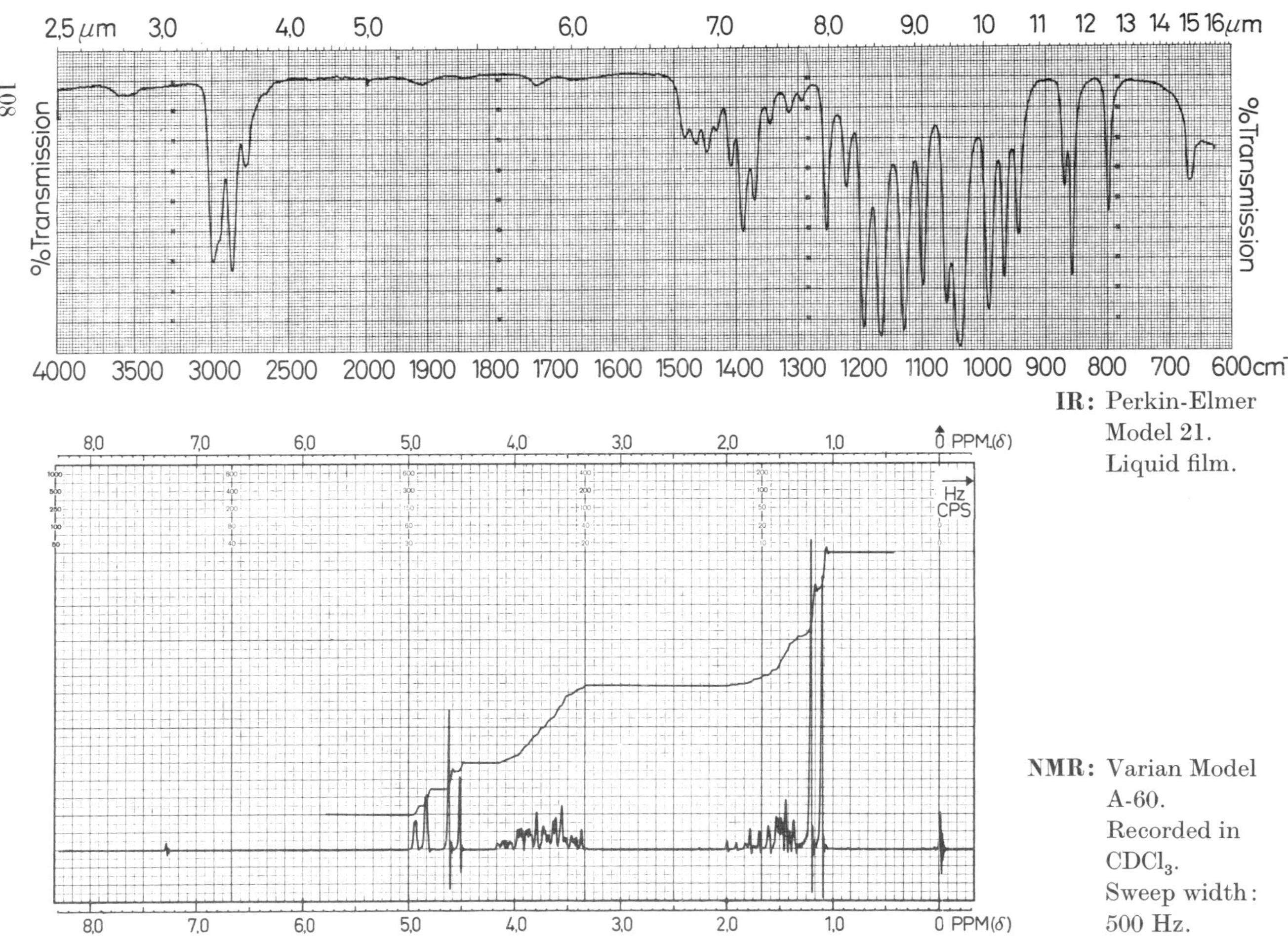

IR: Perkin-Elmer Model 21. Liquid film.

NMR: Varian Model A-60. Recorded in $CDCl_3$. Sweep width: 500 Hz.

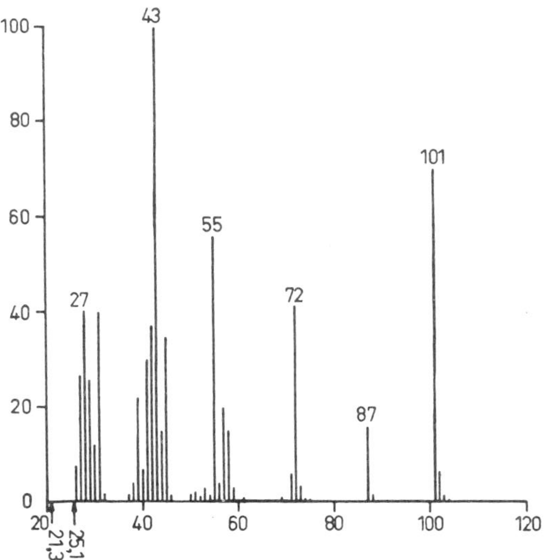

MS: Hitachi Perkin-Elmer Model RMU-6A.

UV: Perkin-Elmer Model 137 UV. Recorded in C_2H_5OH >210 nm none

Problem 8

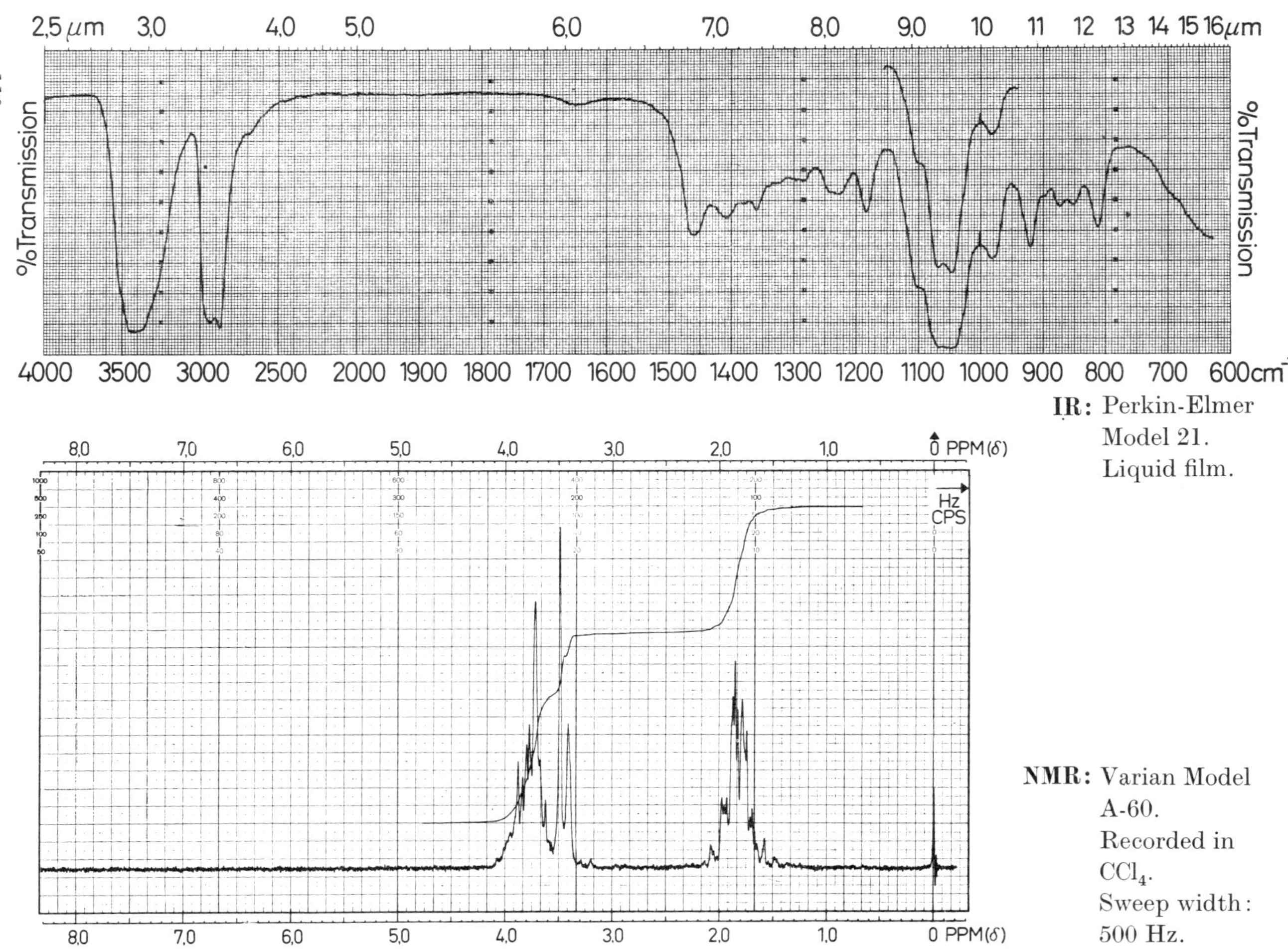

IR: Perkin-Elmer Model 21. Liquid film.

NMR: Varian Model A-60. Recorded in CCl_4. Sweep width: 500 Hz.

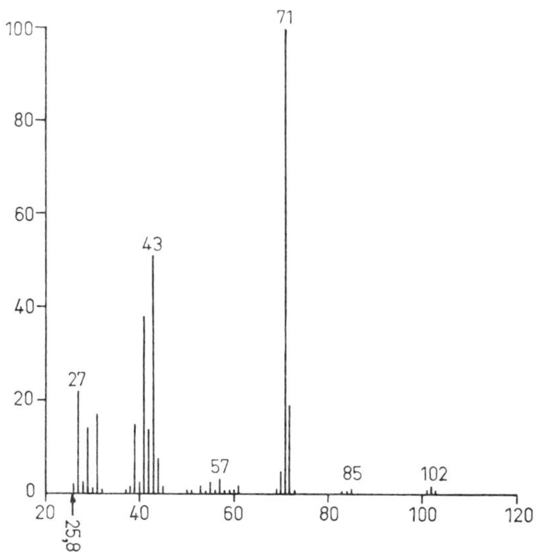

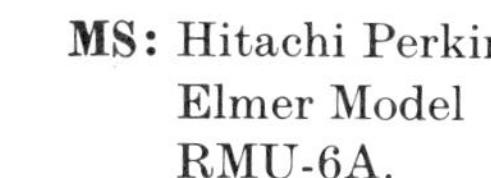

MS: Hitachi Perkin-Elmer Model RMU-6A.

UV: Perkin-Elmer Model 137 UV. Recorded in C_2H_5OH. >210 nm none

Problem 9

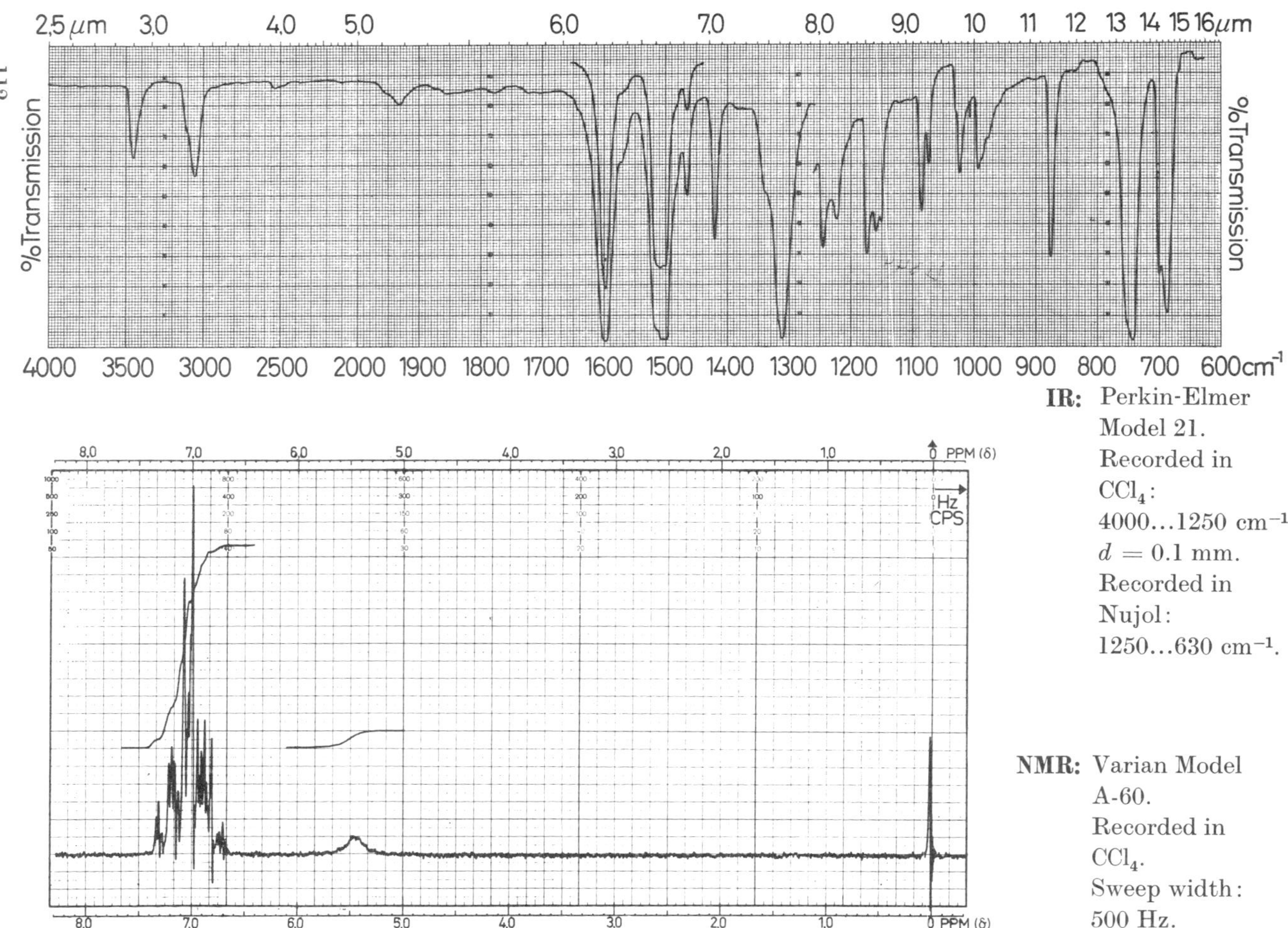

IR: Perkin-Elmer Model 21. Recorded in CCl_4: 4000…1250 cm^{-1} $d = 0.1$ mm. Recorded in Nujol: 1250…630 cm^{-1}.

NMR: Varian Model A-60. Recorded in CCl_4. Sweep width: 500 Hz.

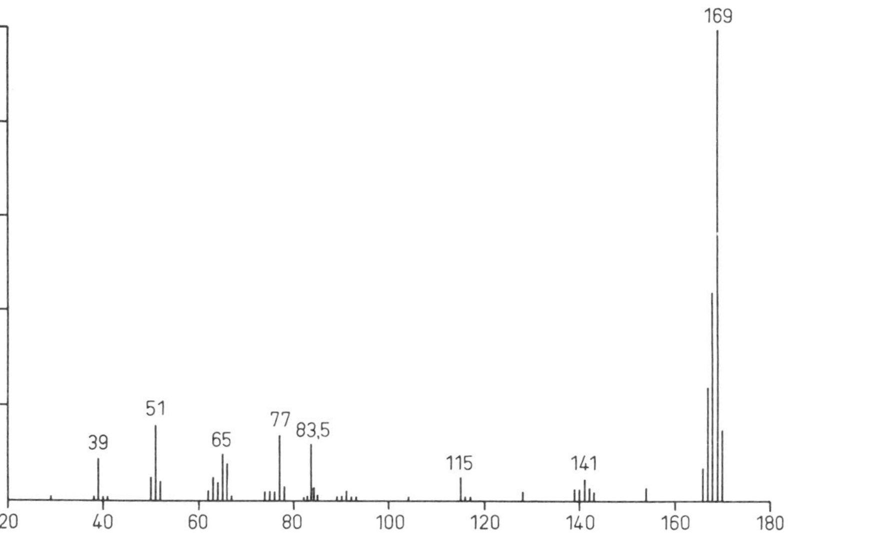

MS: Hitachi Perkin-Elmer Model RMU-6D.

UV: Perkin-Elmer Model 137 UV. Recorded in C_2H_5OH.

λ_{max}	$\log \varepsilon$
285 nm	4.3

Problem 10

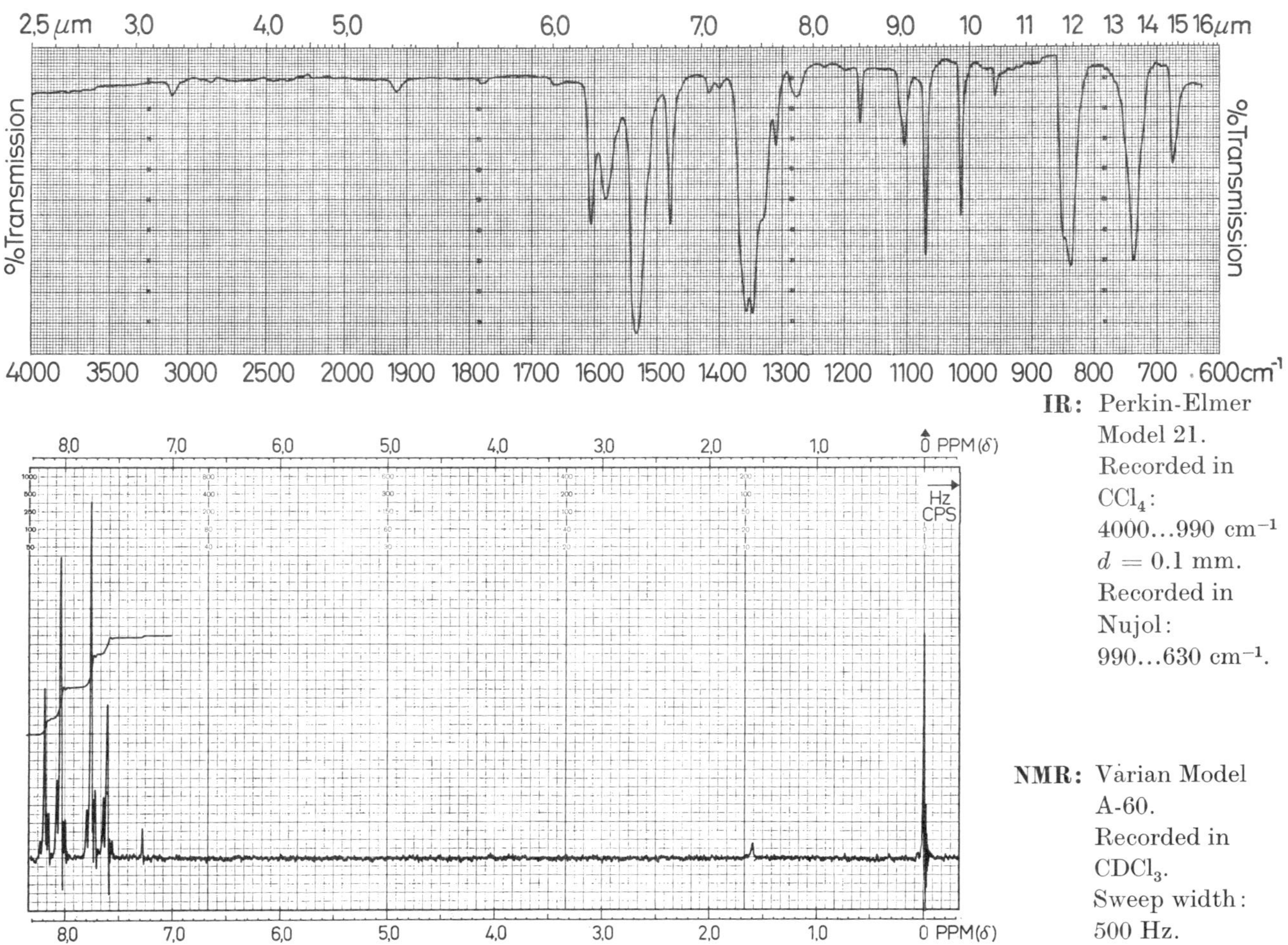

IR: Perkin-Elmer Model 21. Recorded in CCl_4: 4000…990 cm^{-1} $d = 0.1$ mm. Recorded in Nujol: 990…630 cm^{-1}.

NMR: Varian Model A-60. Recorded in $CDCl_3$. Sweep width: 500 Hz.

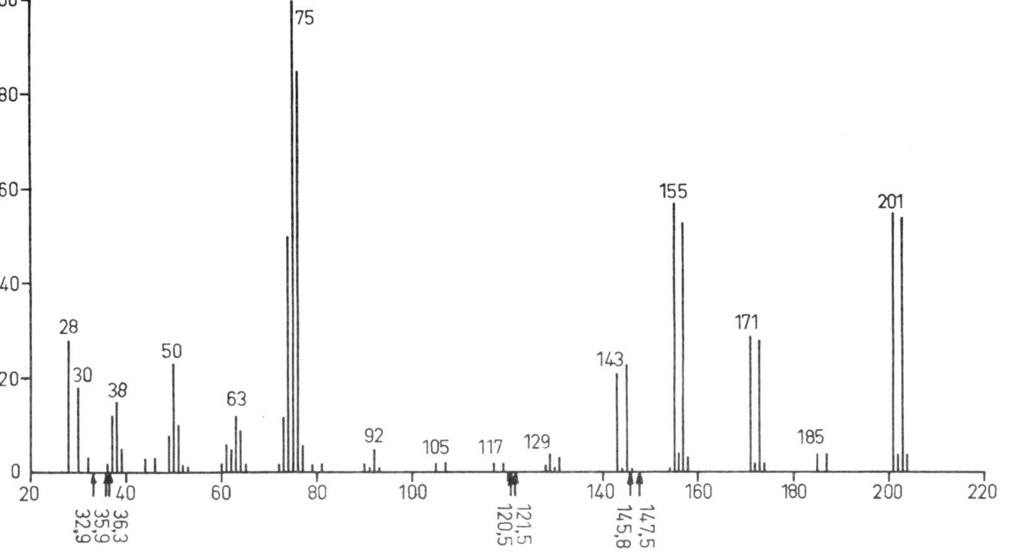

MS: Hitachi Perkin-Elmer Model RMU-6A.

UV: Perkin-Elmer Model 137 UV. Recorded in C_2H_5OH.

λ_{max}	$\log \varepsilon$
270 nm	3.0

Problem 11

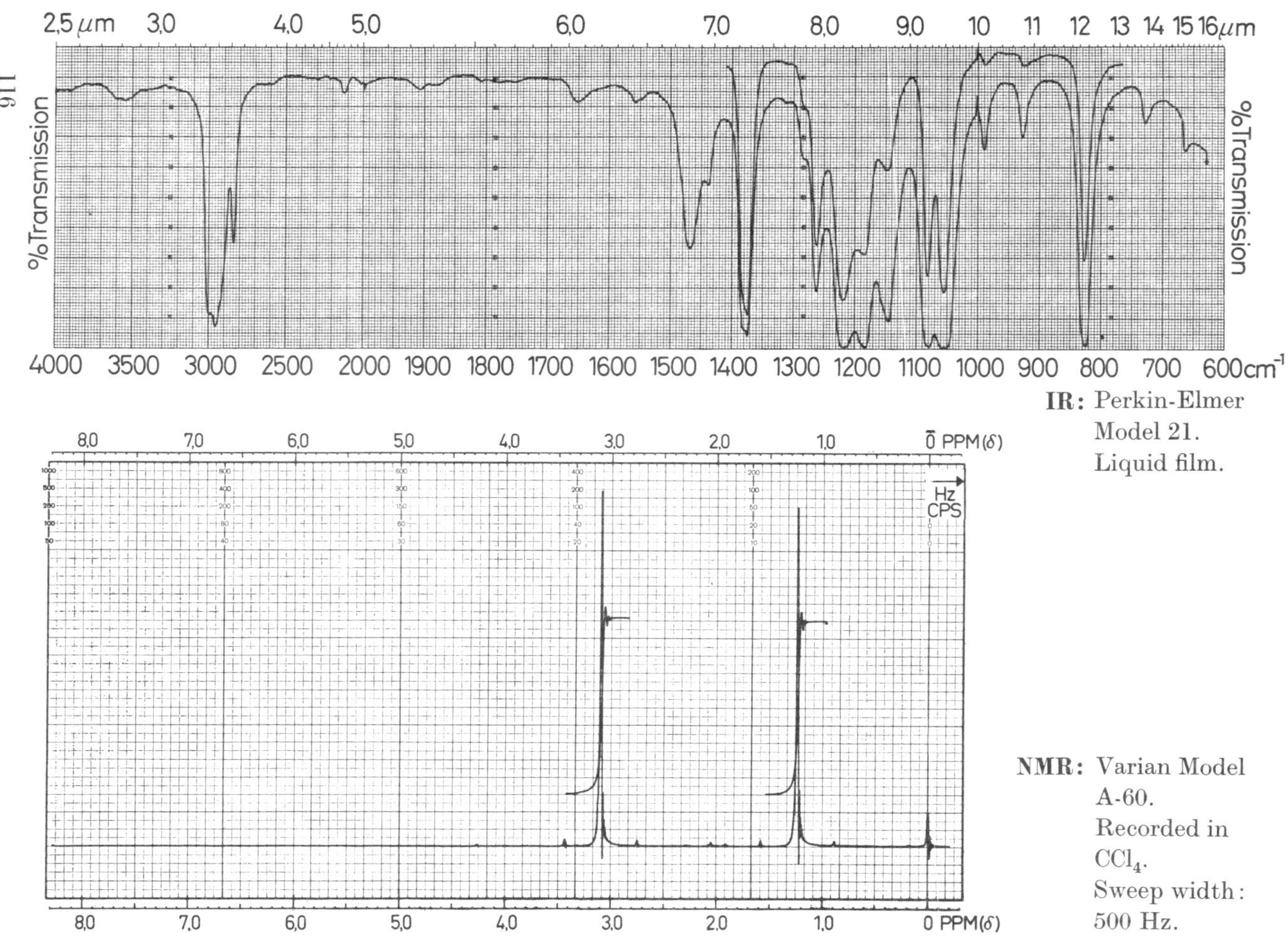

IR: Perkin-Elmer Model 21. Liquid film.

NMR: Varian Model A-60. Recorded in CCl_4. Sweep width: 500 Hz.

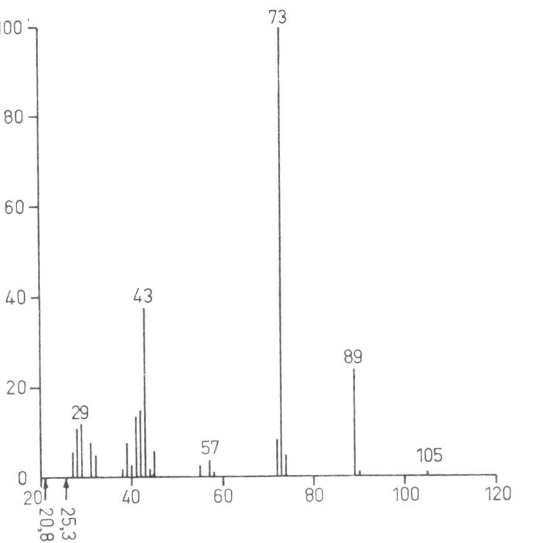

MS: Hitachi Perkin-Elmer Model RMU-6A.

UV: Perkin-Elmer Model 137 UV. Recorded in C_2H_5OH. >210 nm none

Problem 12

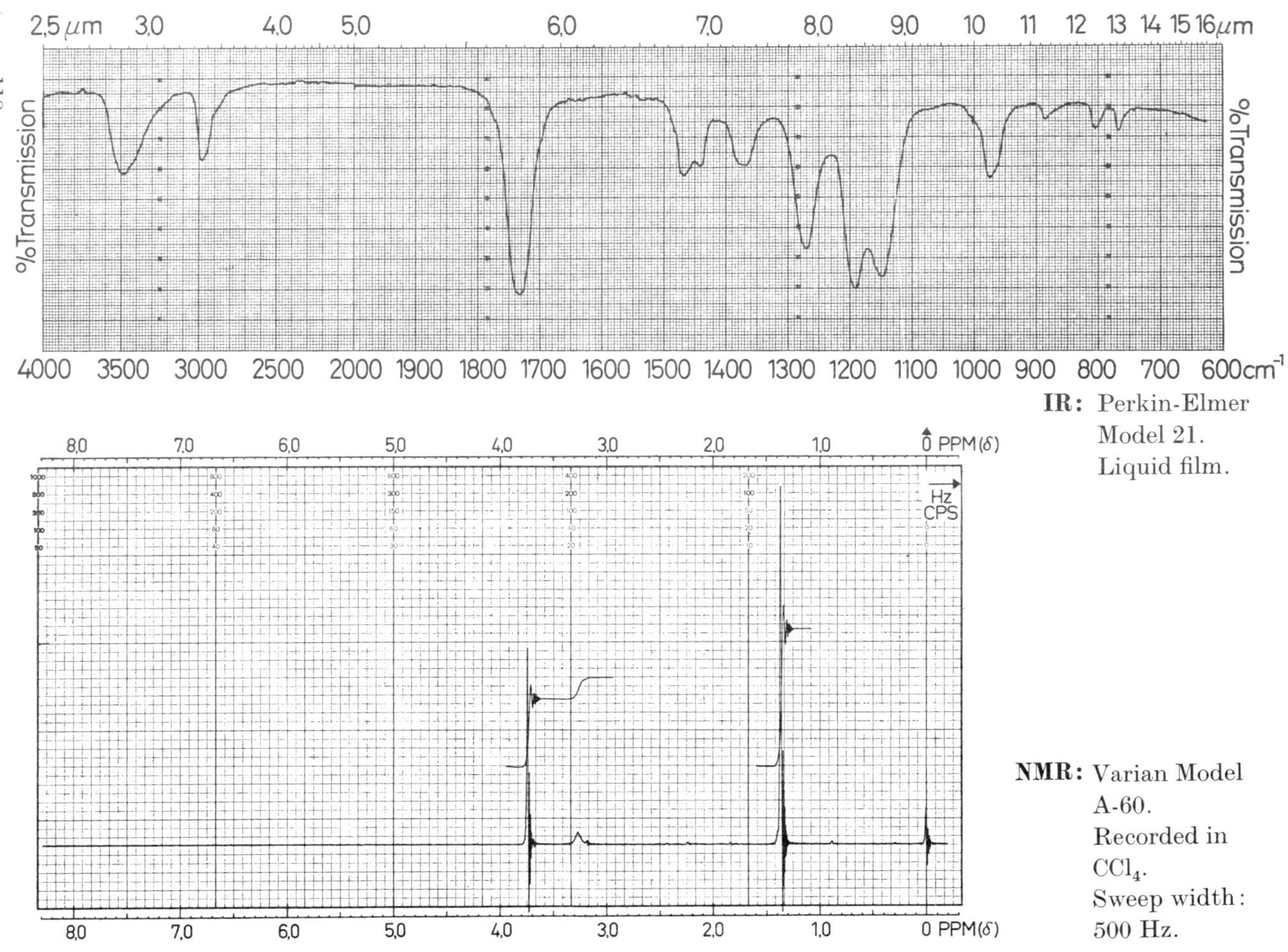

IR: Perkin-Elmer Model 21. Liquid film.

NMR: Varian Model A-60. Recorded in CCl_4. Sweep width: 500 Hz.

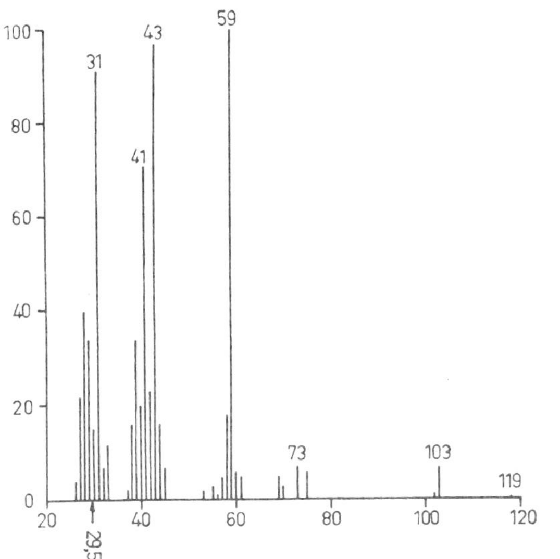

MS: Hitachi Perkin-Elmer Model RMU-6A.

UV: Perkin-Elmer Model 137 UV. Recorded in C_2H_5OH.
>210 nm none

Problem 13

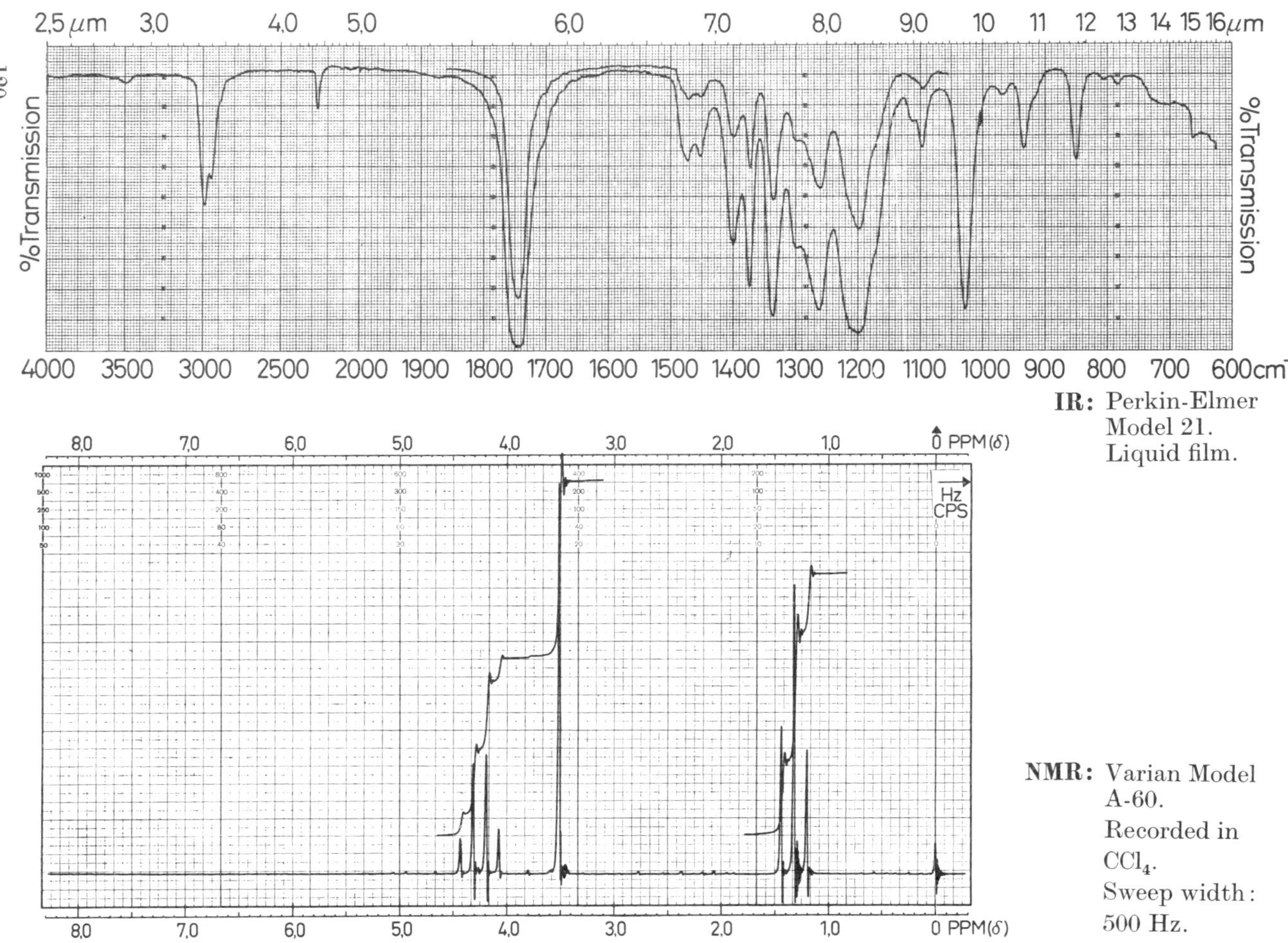

IR: Perkin-Elmer Model 21. Liquid film.

NMR: Varian Model A-60. Recorded in CCl_4. Sweep width: 500 Hz.

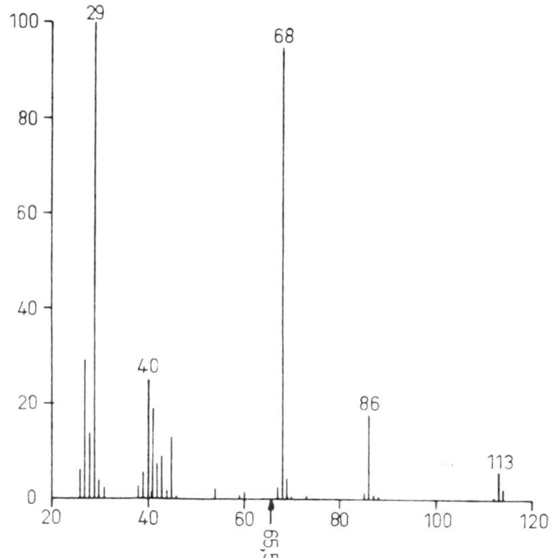

MS: Hitachi Perkin-Elmer Model RMU-6A.

UV: Perkin-Elmer Model 137 UV. Recorded in C_2H_5OH.
>210 nm none

Problem 14

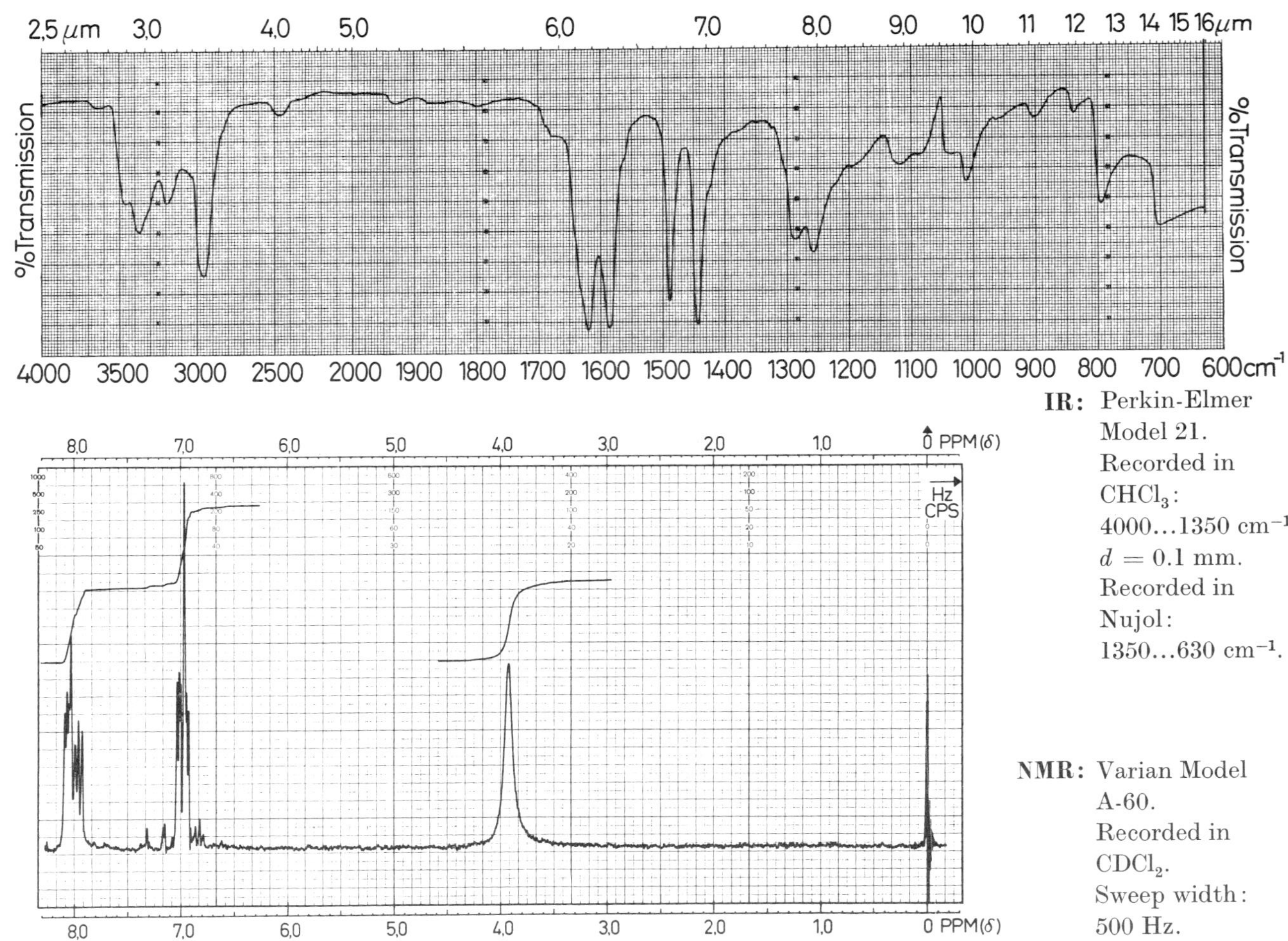

IR: Perkin-Elmer Model 21. Recorded in $CHCl_3$: 4000...1350 cm^{-1} $d = 0.1$ mm. Recorded in Nujol: 1350...630 cm^{-1}.

NMR: Varian Model A-60. Recorded in $CDCl_2$. Sweep width: 500 Hz.

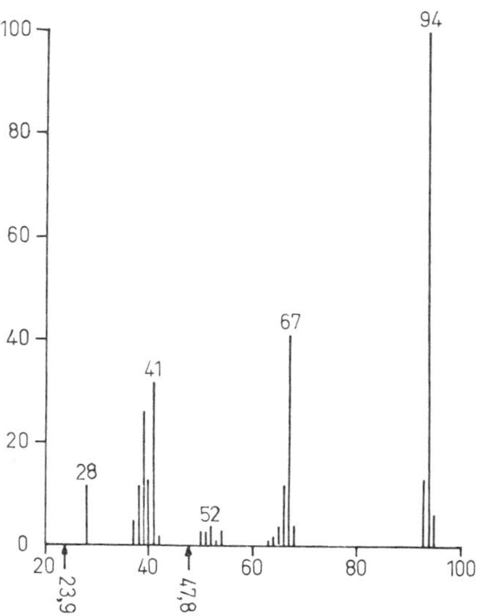

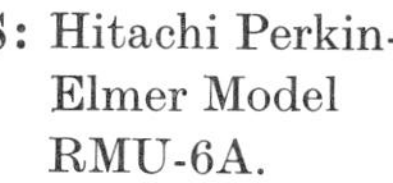

MS: Hitachi Perkin-Elmer Model RMU-6A.

UV: Perkin-Elmer Model 137 UV. Recorded in C_2H_5OH.

λ_{max}	log ε
242 nm	4.0
301 nm	3.5

Problem 15

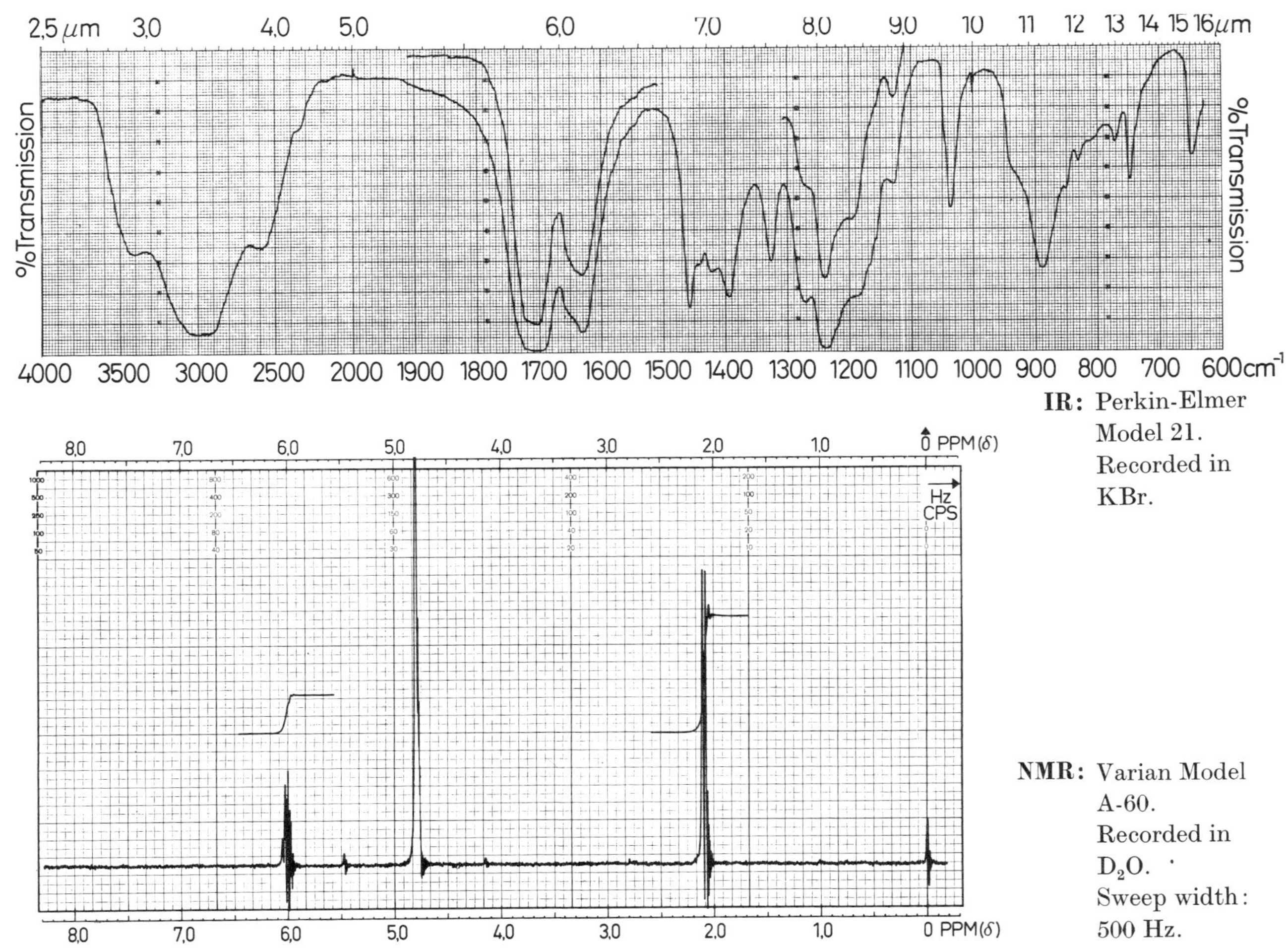

IR: Perkin-Elmer Model 21. Recorded in KBr.

NMR: Varian Model A-60. Recorded in D_2O. Sweep width: 500 Hz.

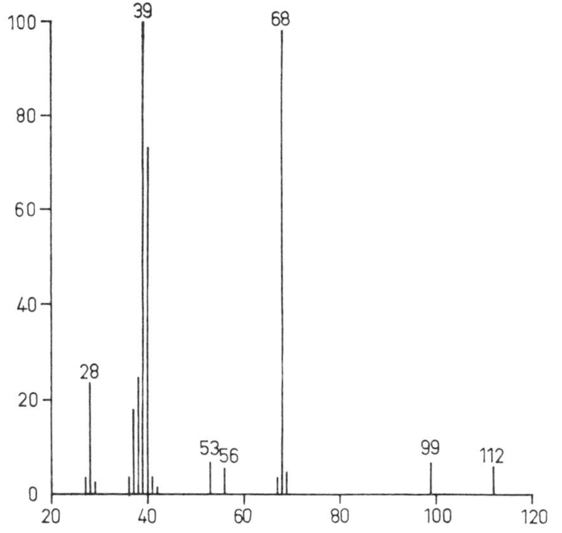

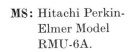

MS: Hitachi Perkin-Elmer Model RMU-6A.

UV: Perkin-Elmer Model 137 UV. Recorded in C_2H_5OH.

λ_{max}	log ε
218 nm	2.9

Problem 16

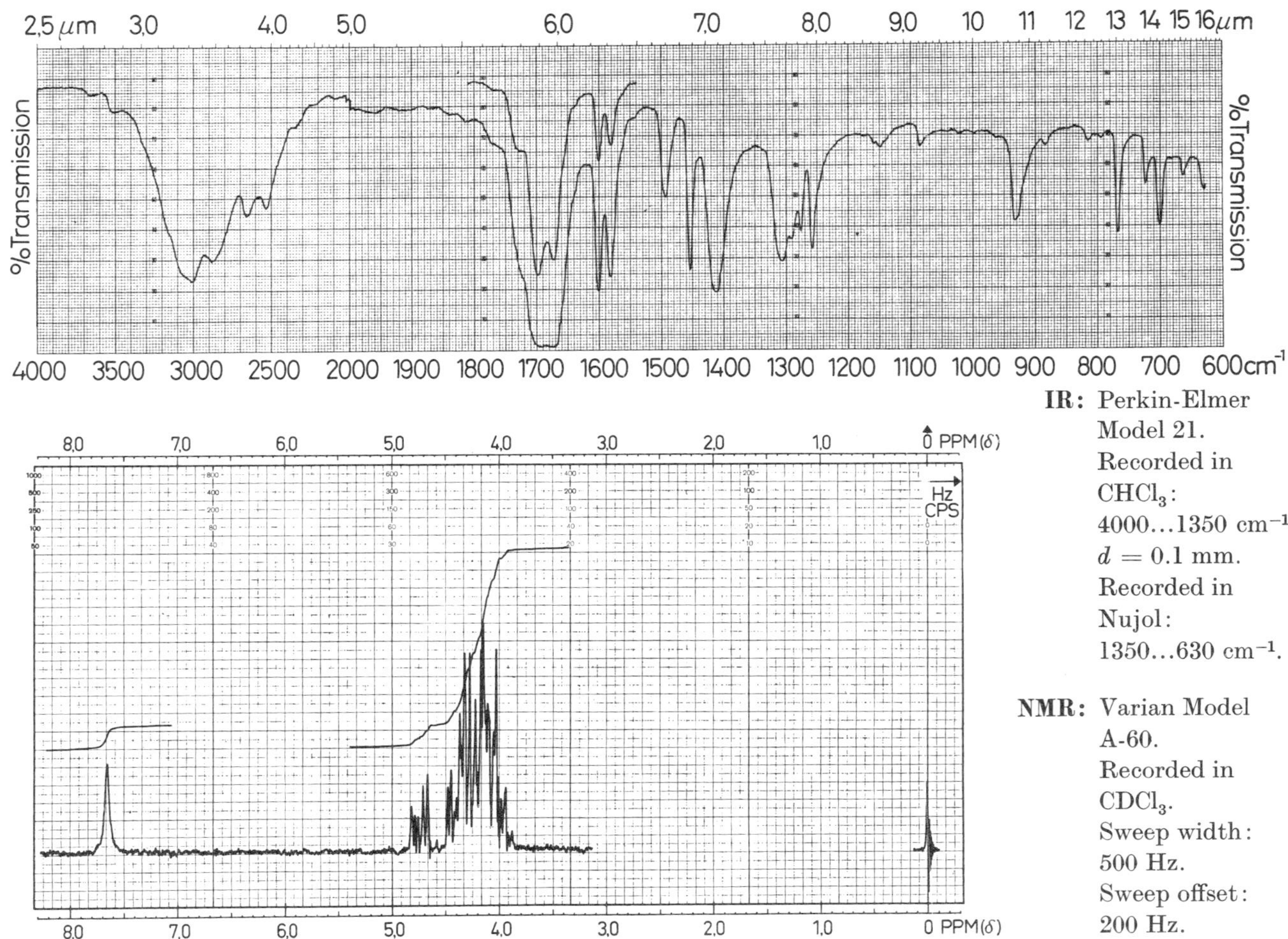

IR: Perkin-Elmer Model 21. Recorded in $CHCl_3$: 4000…1350 cm^{-1} $d = 0.1$ mm. Recorded in Nujol: 1350…630 cm^{-1}.

NMR: Varian Model A-60. Recorded in $CDCl_3$. Sweep width: 500 Hz. Sweep offset: 200 Hz.

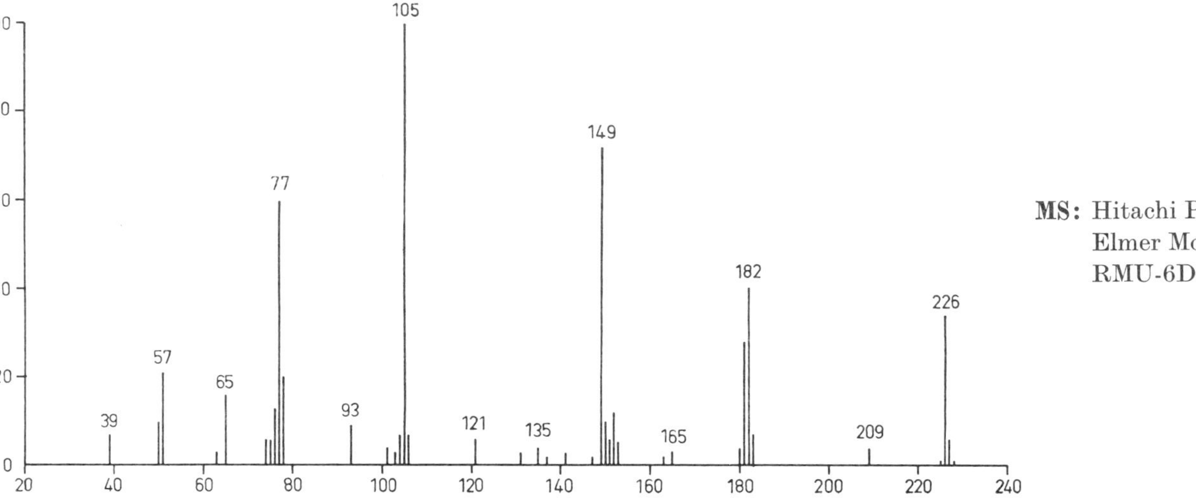

MS: Hitachi Perkin-Elmer Model RMU-6D.

UV: Perkin-Elmer Model 137 UV. Recorded in C_2H_5OH.

λ_{max}	$\log \varepsilon$
246 nm	4.1
280 nm	3.5

Problem 17

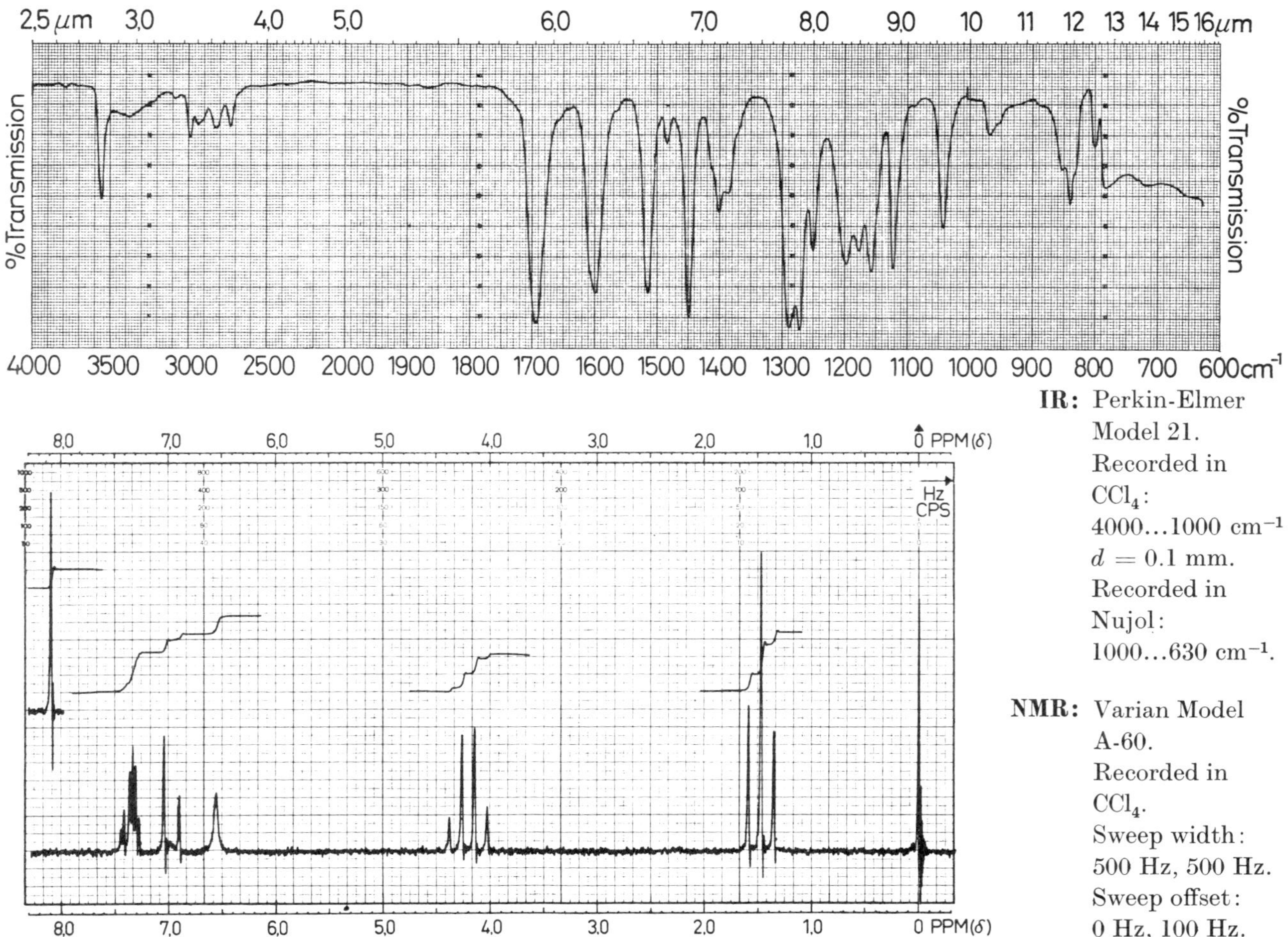

IR: Perkin-Elmer Model 21. Recorded in CCl_4: 4000...1000 cm^{-1} $d = 0.1$ mm. Recorded in Nujol: 1000...630 cm^{-1}.

NMR: Varian Model A-60. Recorded in CCl_4. Sweep width: 500 Hz, 500 Hz. Sweep offset: 0 Hz, 100 Hz.

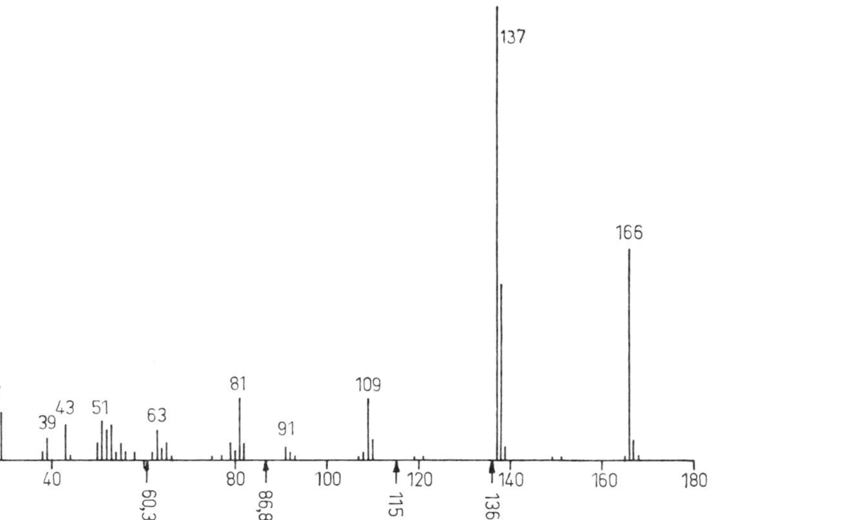

MS: Hitachi Perkin-Elmer Model RMU-6D.

UV: Perkin-Elmer Model 137 UV. Recorded in C_2H_5OH.

λ_{max}	$\log \varepsilon$
233 nm	4.6
280 nm	4.3
310 nm	4.3

Problem 18

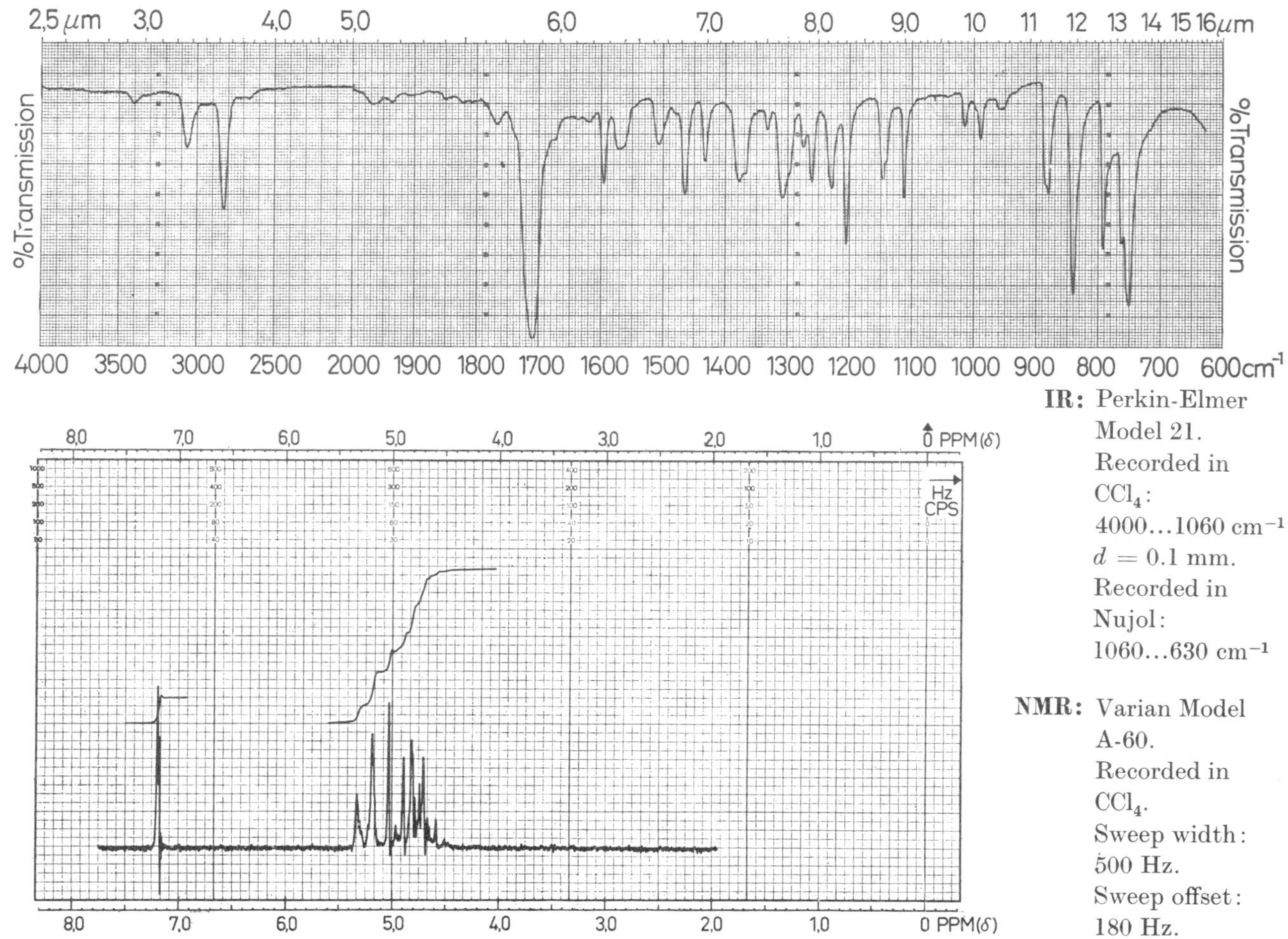

IR: Perkin-Elmer Model 21. Recorded in CCl_4: 4000...1060 cm^{-1} $d = 0.1$ mm. Recorded in Nujol: 1060...630 cm^{-1}

NMR: Varian Model A-60. Recorded in CCl_4. Sweep width: 500 Hz. Sweep offset: 180 Hz.

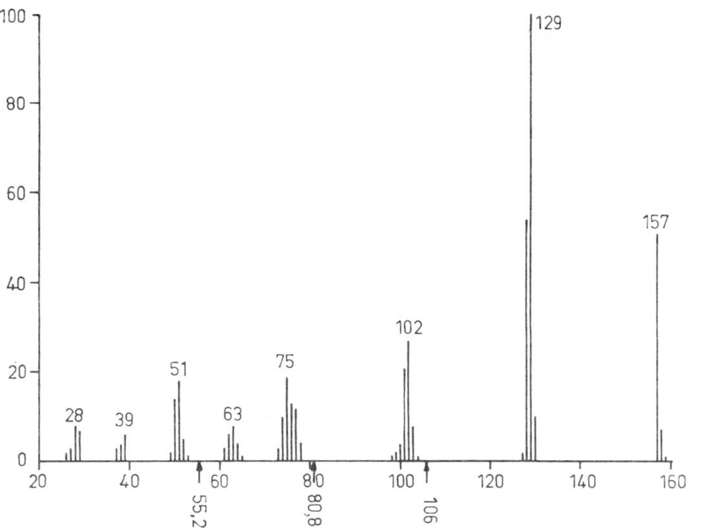

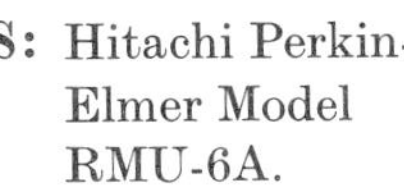

MS: Hitachi Perkin-Elmer Model RMU-6A.

UV: Perkin-Elmer Model 137 UV. Recorded in C_2H_5OH.

λ_{max}	$\log \varepsilon$
229 nm s	4.4
233 nm	4.5
248 nm	4.1
290 nm	3.6
302 nm	3.6
315 nm	3.5

Problem 19

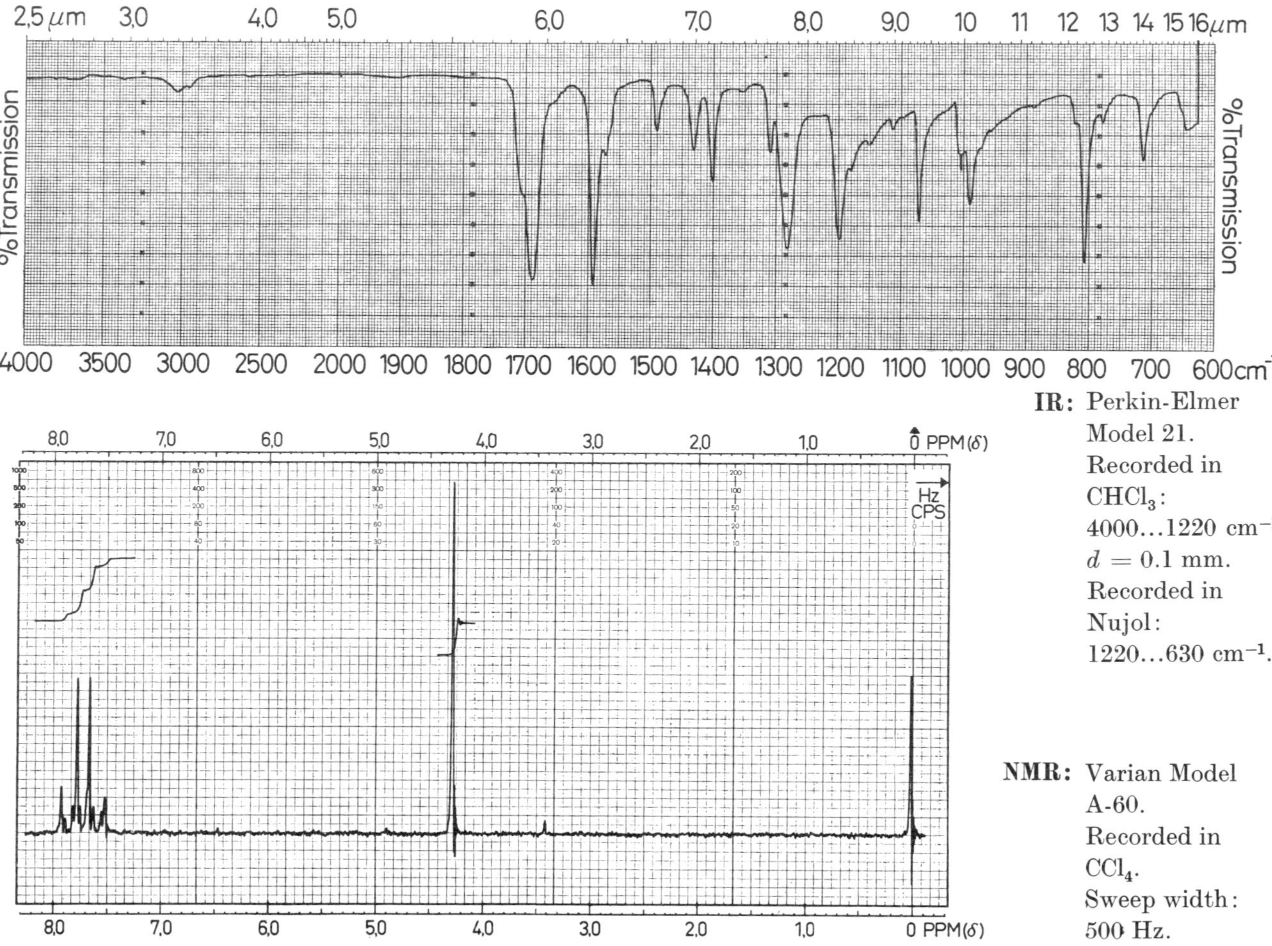

IR: Perkin-Elmer Model 21. Recorded in $CHCl_3$: 4000…1220 cm^{-1} $d = 0.1$ mm. Recorded in Nujol: 1220…630 cm^{-1}.

NMR: Varian Model A-60. Recorded in CCl_4. Sweep width: 500 Hz.

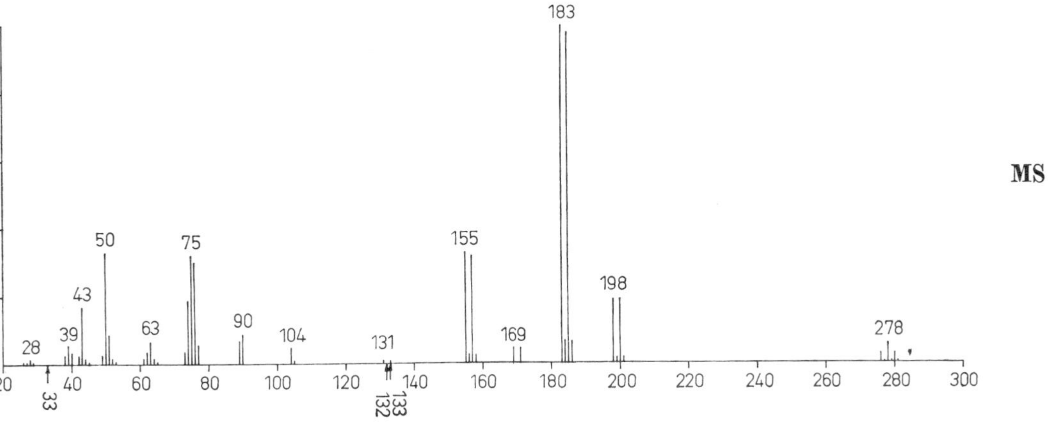

MS: Hitachi Perkin-Elmer Model RMU-6A.

UV: Perkin-Elmer Model 137 UV. Recorded in C_2H_5OH.

λ_{max}	log ε
263 nm	4.0

Problem 20

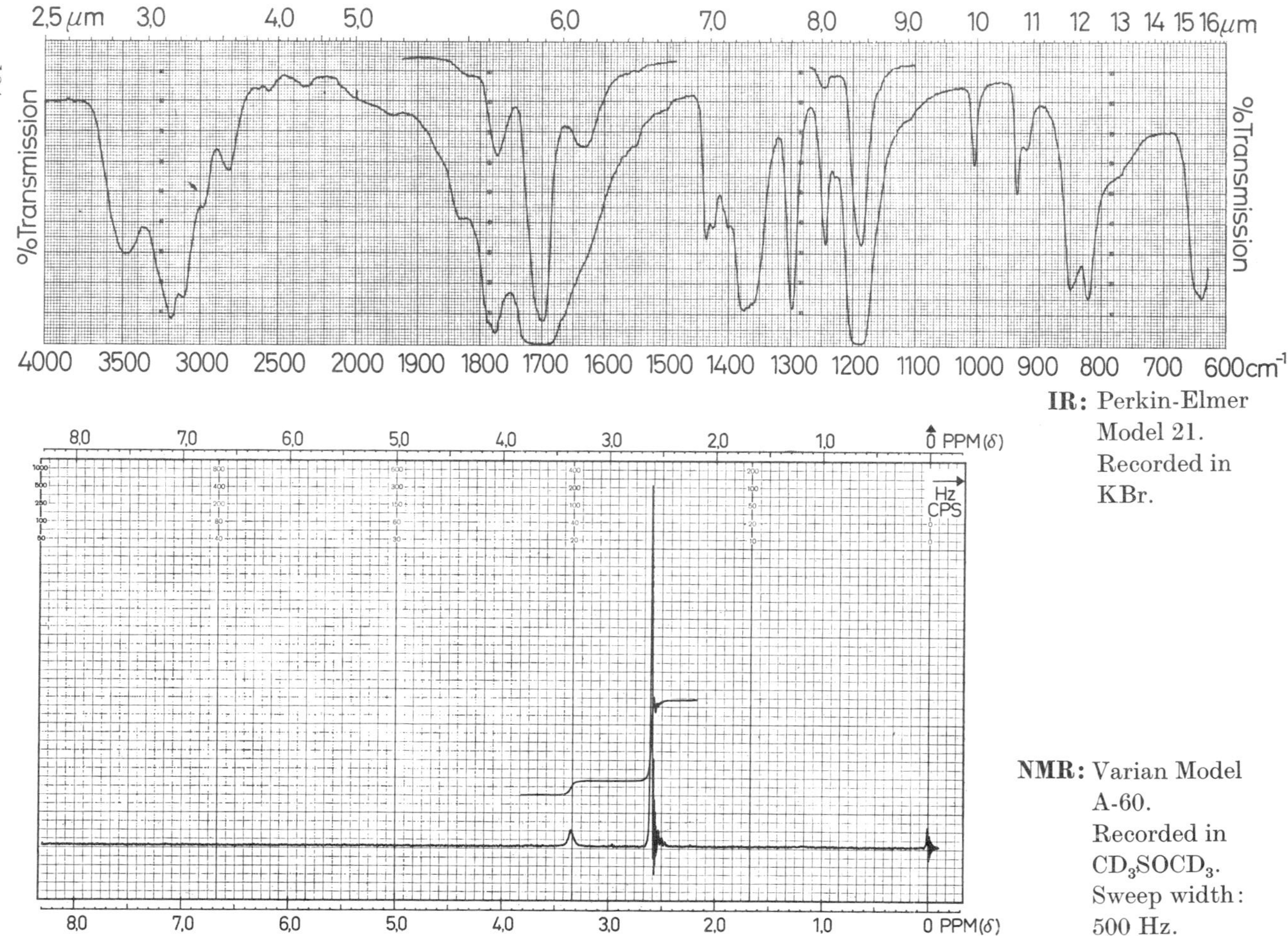

IR: Perkin-Elmer Model 21. Recorded in KBr.

NMR: Varian Model A-60. Recorded in CD_3SOCD_3. Sweep width: 500 Hz.

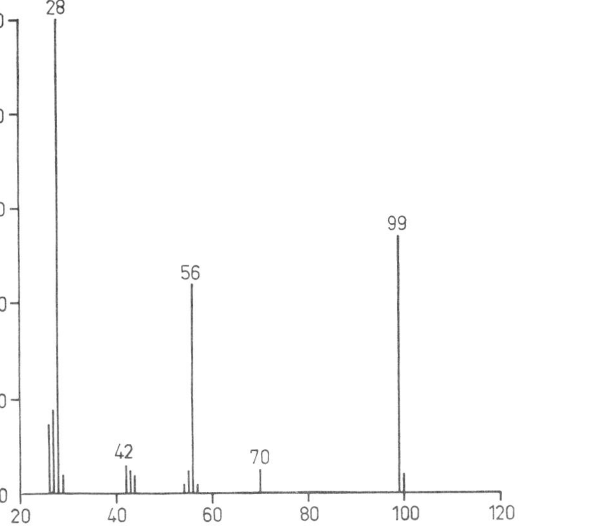

MS: Hitachi Perkin-Elmer Model RMU-6D.

UV: Perkin-Elmer Model 137 UV. Recorded in C_2H_5OH.

λ_{max}	$\log \varepsilon$
221 nm s	2.2
245 nm s	1.9

Problem 21

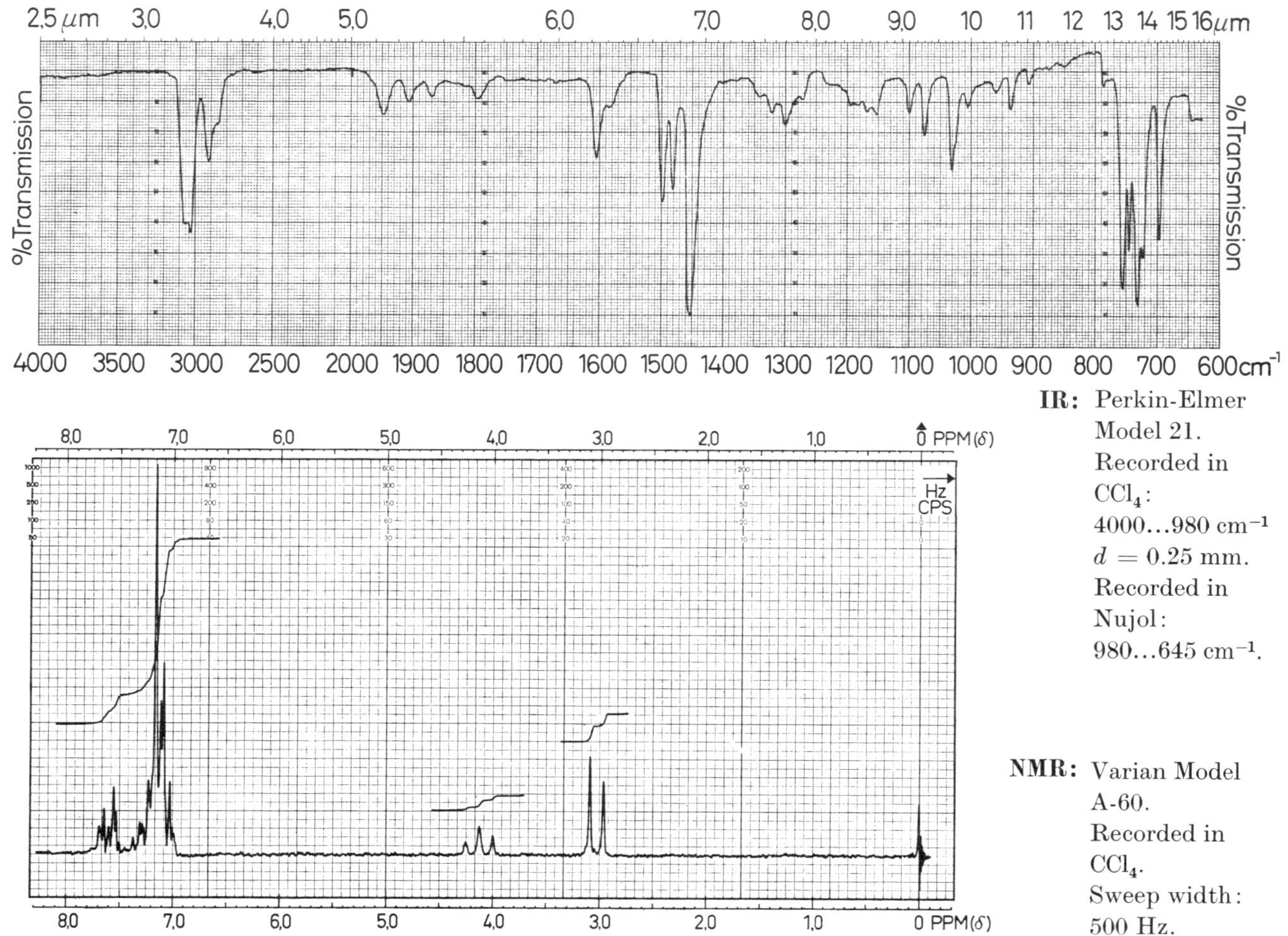

IR: Perkin-Elmer Model 21. Recorded in CCl_4: 4000...980 cm^{-1} $d = 0.25$ mm. Recorded in Nujol: 980...645 cm^{-1}.

NMR: Varian Model A-60. Recorded in CCl_4. Sweep width: 500 Hz.

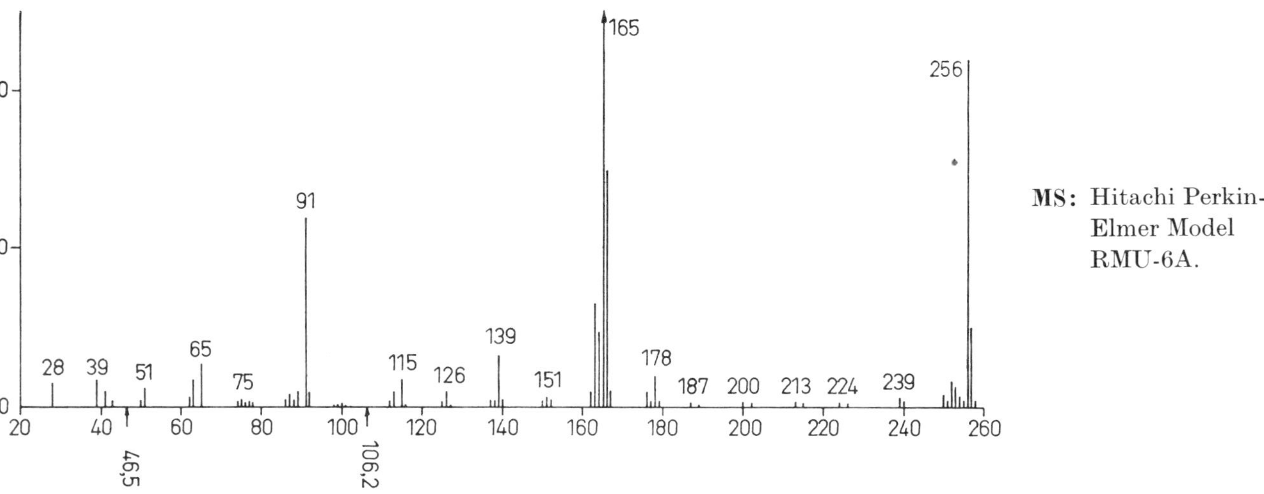

MS: Hitachi Perkin-Elmer Model RMU-6A.

UV: Perkin-Elmer Model 137 UV. Recorded in C_2H_5OH.

λ_{max}	log ε
230 nm	4.0
246 nm	4.2
291 nm	3.8
301 nm	3.9

Problem 22

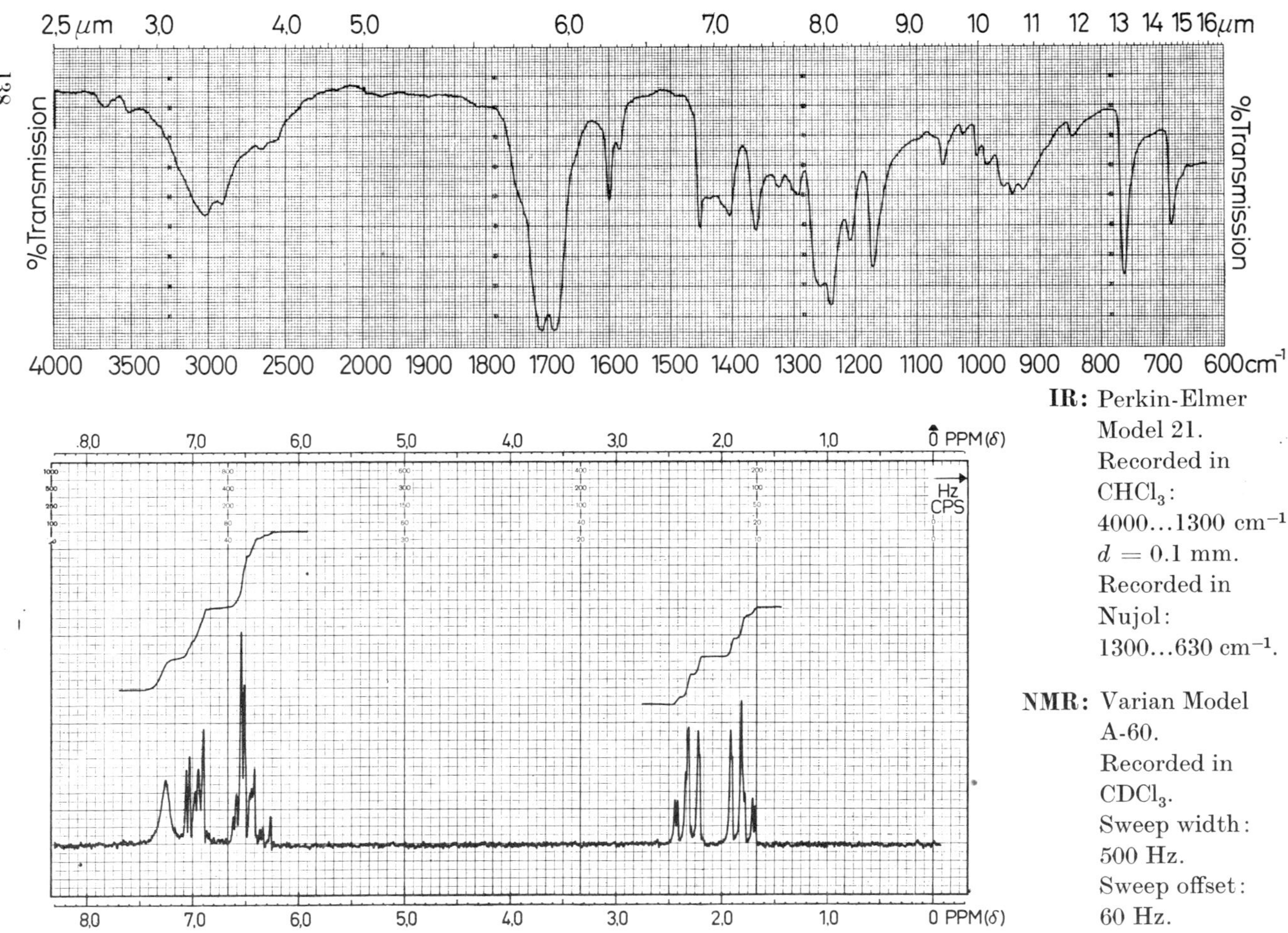

IR: Perkin-Elmer Model 21. Recorded in $CHCl_3$: 4000…1300 cm^{-1} $d = 0.1$ mm. Recorded in Nujol: 1300…630 cm^{-1}.

NMR: Varian Model A-60. Recorded in $CDCl_3$. Sweep width: 500 Hz. Sweep offset: 60 Hz.

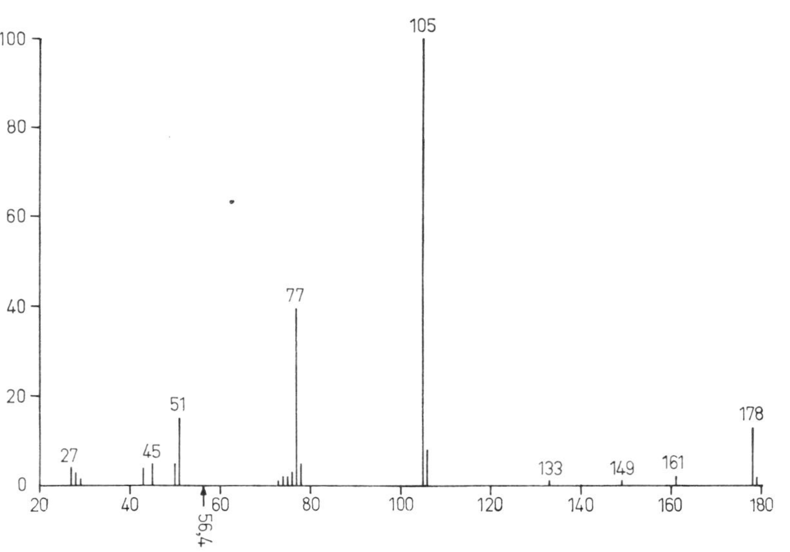

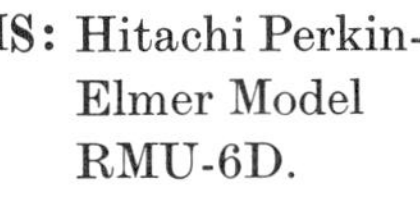

MS: Hitachi Perkin-Elmer Model RMU-6D.

UV: Perkin-Elmer Model 137 UV. Recorded in C_2H_5OH.

λ_{max}	$\log \varepsilon$
242 nm	4.0
280 nm	2.9

Problem 23

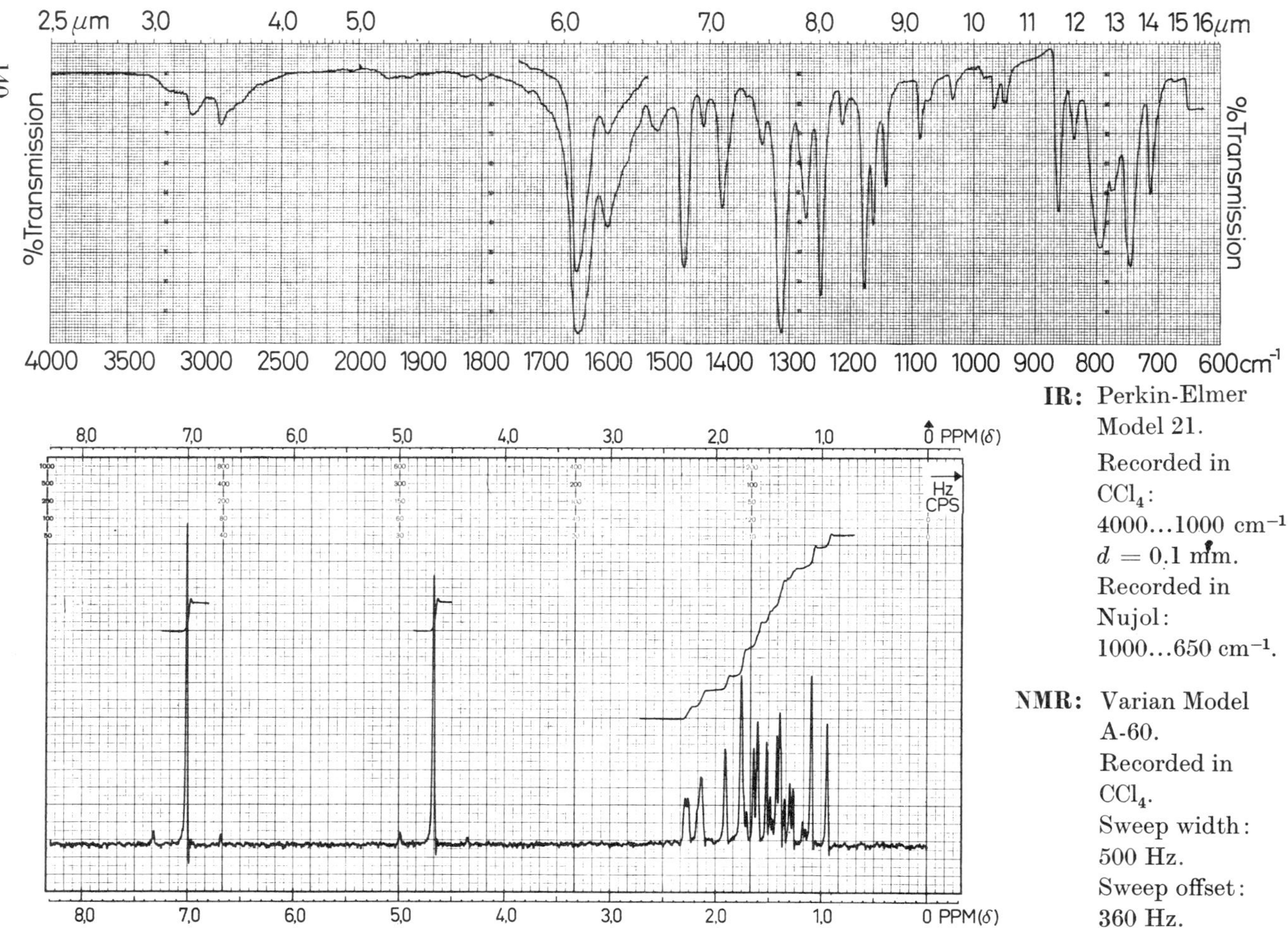

IR: Perkin-Elmer Model 21.
Recorded in CCl_4: 4000...1000 cm^{-1} $d = 0.1$ mm.
Recorded in Nujol: 1000...650 cm^{-1}.

NMR: Varian Model A-60.
Recorded in CCl_4.
Sweep width: 500 Hz.
Sweep offset: 360 Hz.

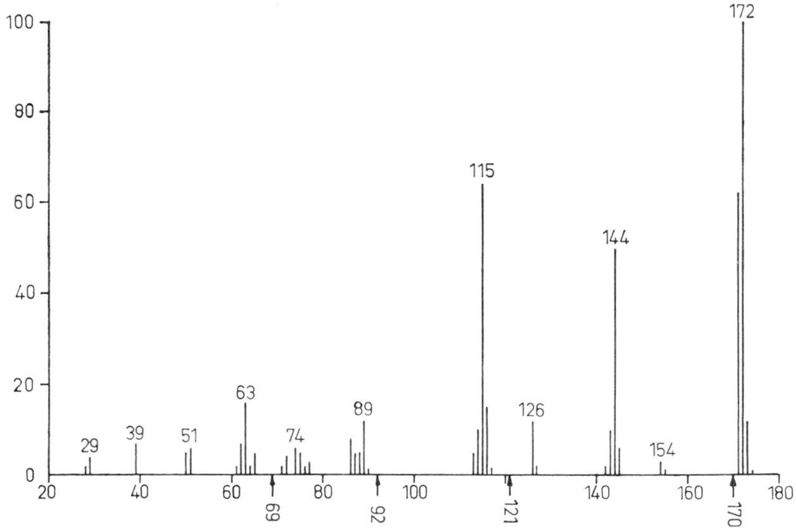

MS: Hitachi Perkin-Elmer Model RMU-6A.

UV: Perkin-Elmer Model 137 UV. Recorded in C_2H_5OH.

λ_{max}	log ε
222 nm	4.6
238 nm	4.1
254 nm s	3.9
318 nm	3.8
359 nm	3.7

Problem 24

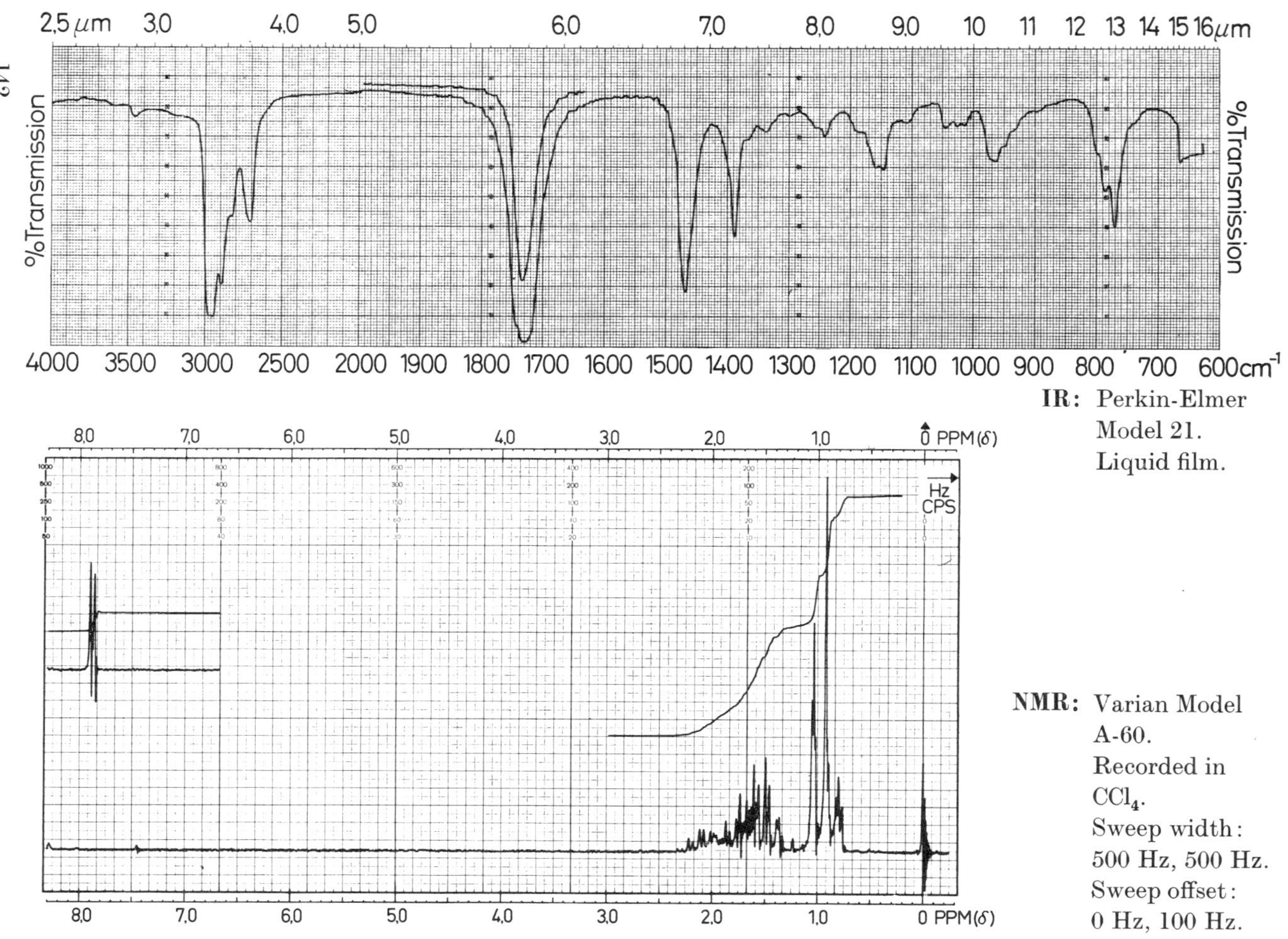

IR: Perkin-Elmer Model 21. Liquid film.

NMR: Varian Model A-60. Recorded in CCl_4. Sweep width: 500 Hz, 500 Hz. Sweep offset: 0 Hz, 100 Hz.

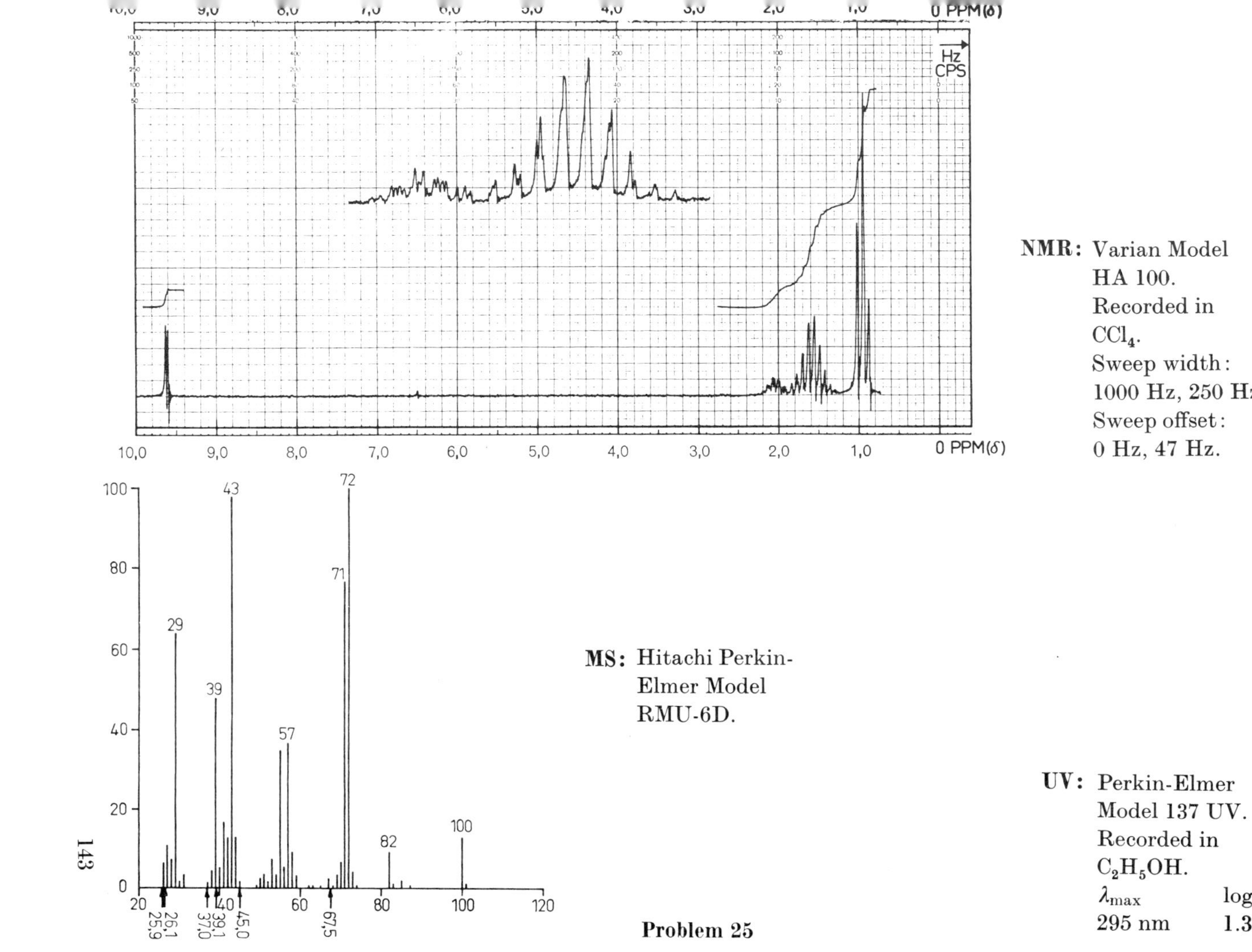

NMR: Varian Model HA 100. Recorded in CCl_4. Sweep width: 1000 Hz, 250 Hz. Sweep offset: 0 Hz, 47 Hz.

MS: Hitachi Perkin-Elmer Model RMU-6D.

UV: Perkin-Elmer Model 137 UV. Recorded in C_2H_5OH.

λ_{max}	$\log \varepsilon$
295 nm	1.3

Problem 25

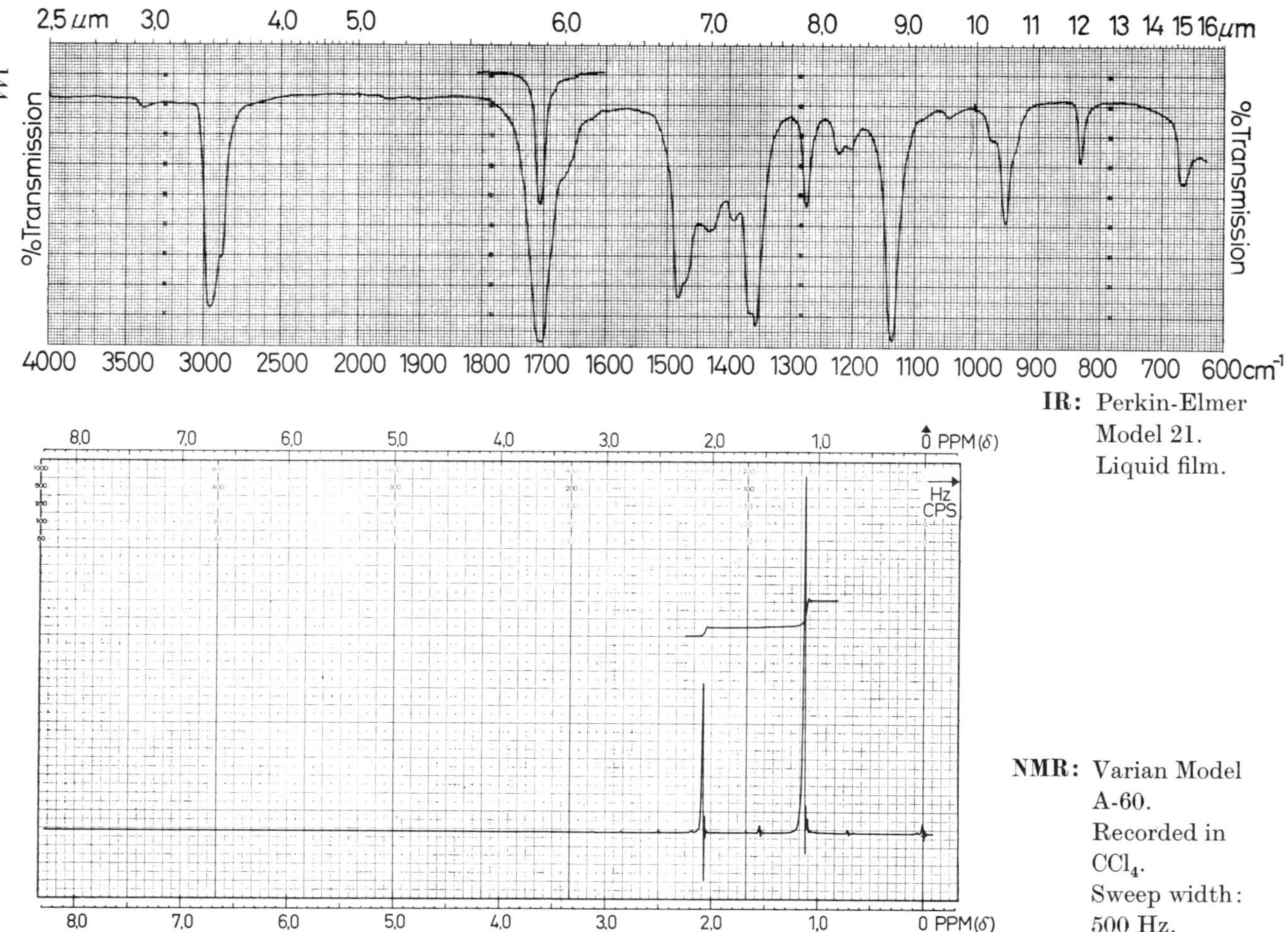

IR: Perkin-Elmer Model 21. Liquid film.

NMR: Varian Model A-60. Recorded in CCl_4. Sweep width: 500 Hz.

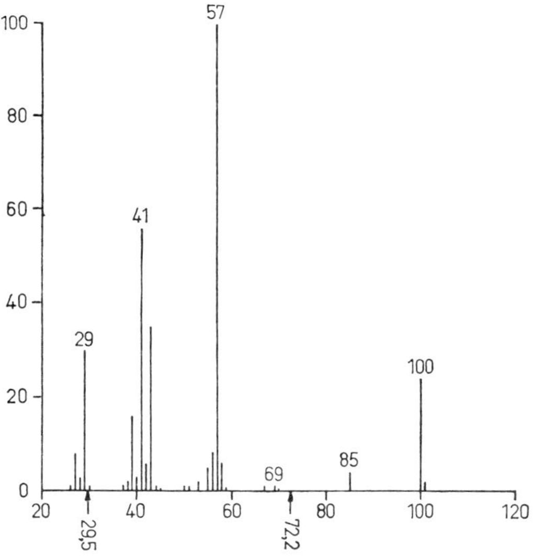

MS: Hitachi Perkin-Elmer Model RMU-6A.

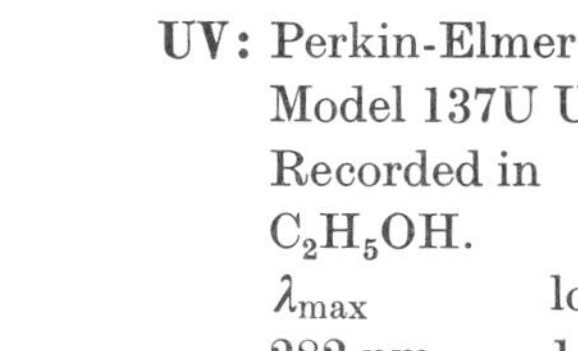

UV: Perkin-Elmer Model 137U UV. Recorded in C_2H_5OH.

λ_{max}	$\log \varepsilon$
282 nm	1.4

Problem 26

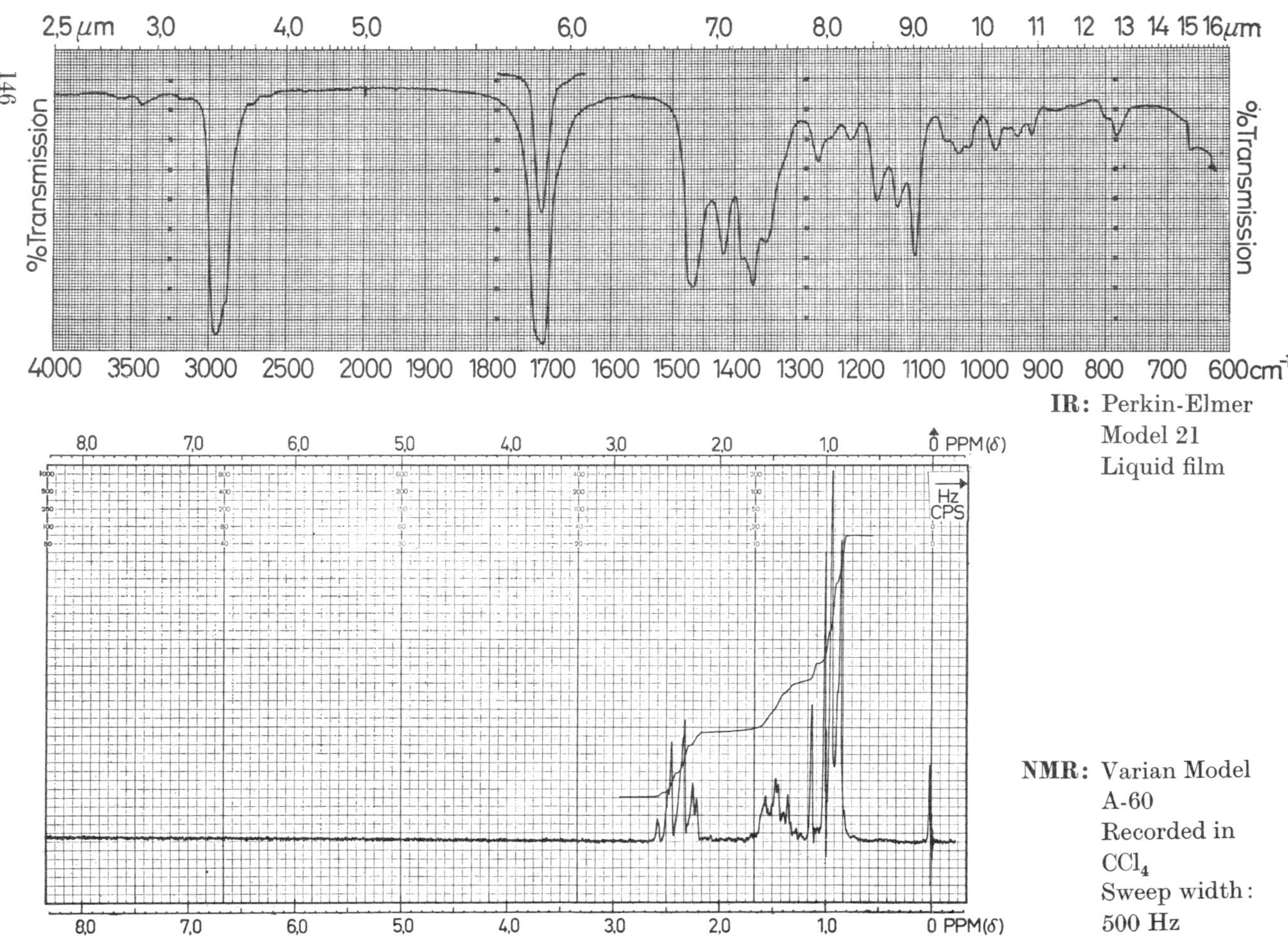

IR: Perkin-Elmer
Model 21
Liquid film

NMR: Varian Model
A-60
Recorded in
CCl_4
Sweep width:
500 Hz

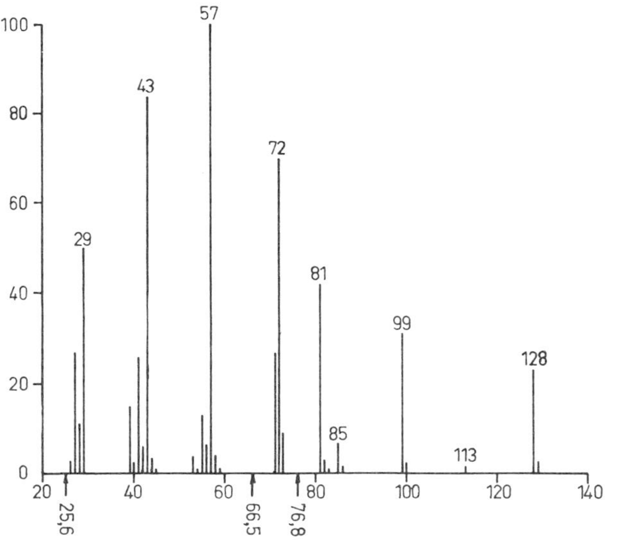

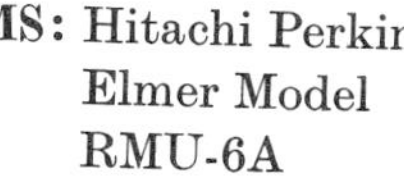

MS: Hitachi Perkin-Elmer Model RMU-6A

UV: Perkin-Elmer Model 137 UV Recorded in C_2H_5OH

λ_{max}	log
278 nm	1.5

Problem 27

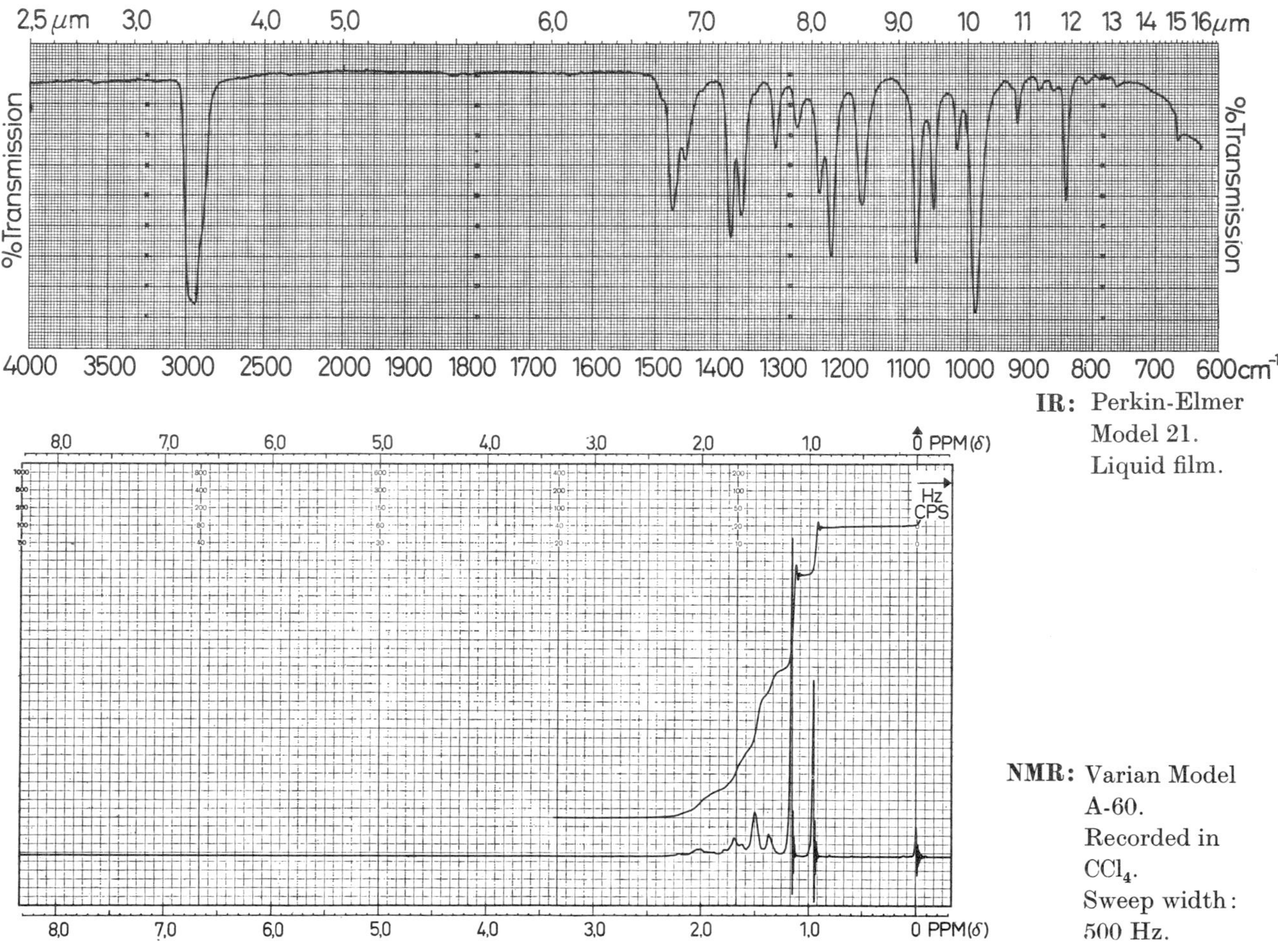

IR: Perkin-Elmer Model 21. Liquid film.

NMR: Varian Model A-60. Recorded in CCl_4. Sweep width: 500 Hz.

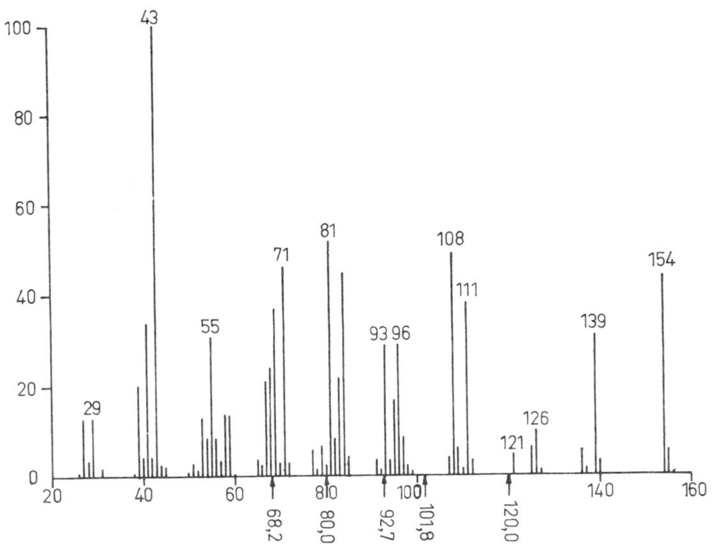

MS: Hitachi Perkin-Elmer Model RMU-6A.

UV: Perkin-Elmer Model 137 UV. Recorded in C_2H_5OH. >210 nm none

Problem 28

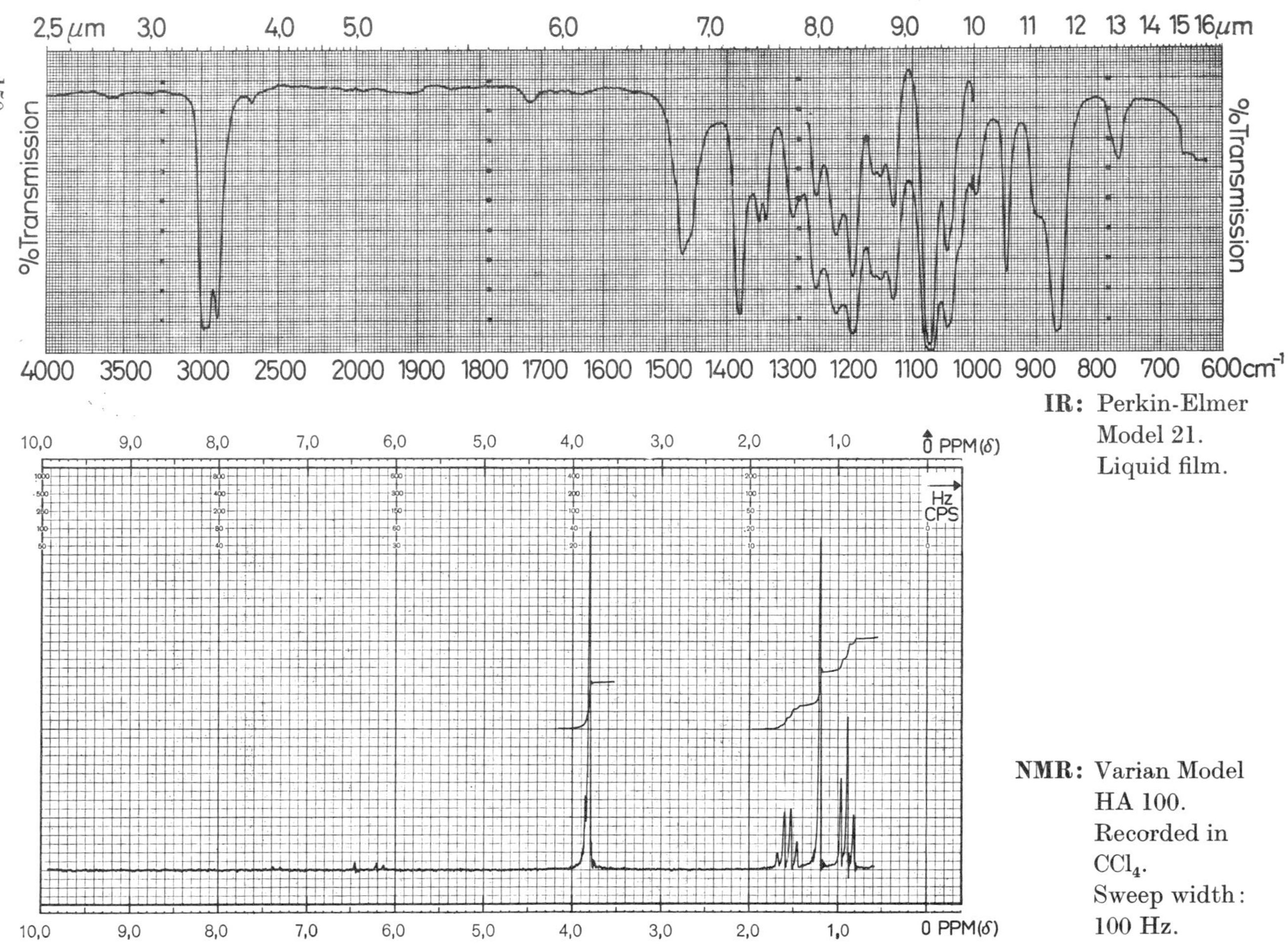

IR: Perkin-Elmer Model 21. Liquid film.

NMR: Varian Model HA 100. Recorded in CCl_4. Sweep width: 100 Hz.

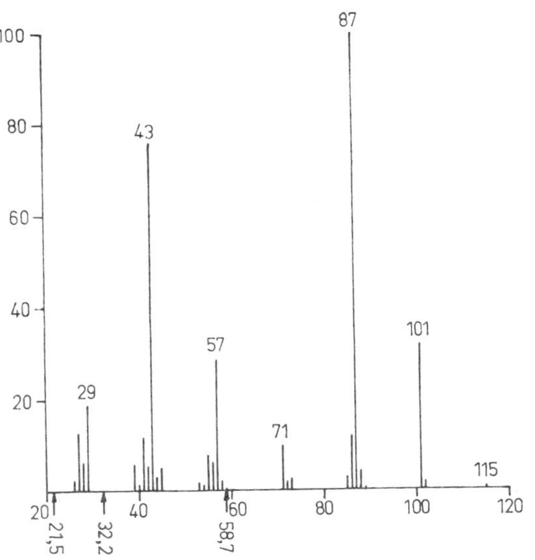

MS: Hitachi Perkin-Elmer Model RMU-6A.

UV: Perkin-Elmer Model 137 UV. Recorded in C_2H_5OH. >210 nm none

Problem 29

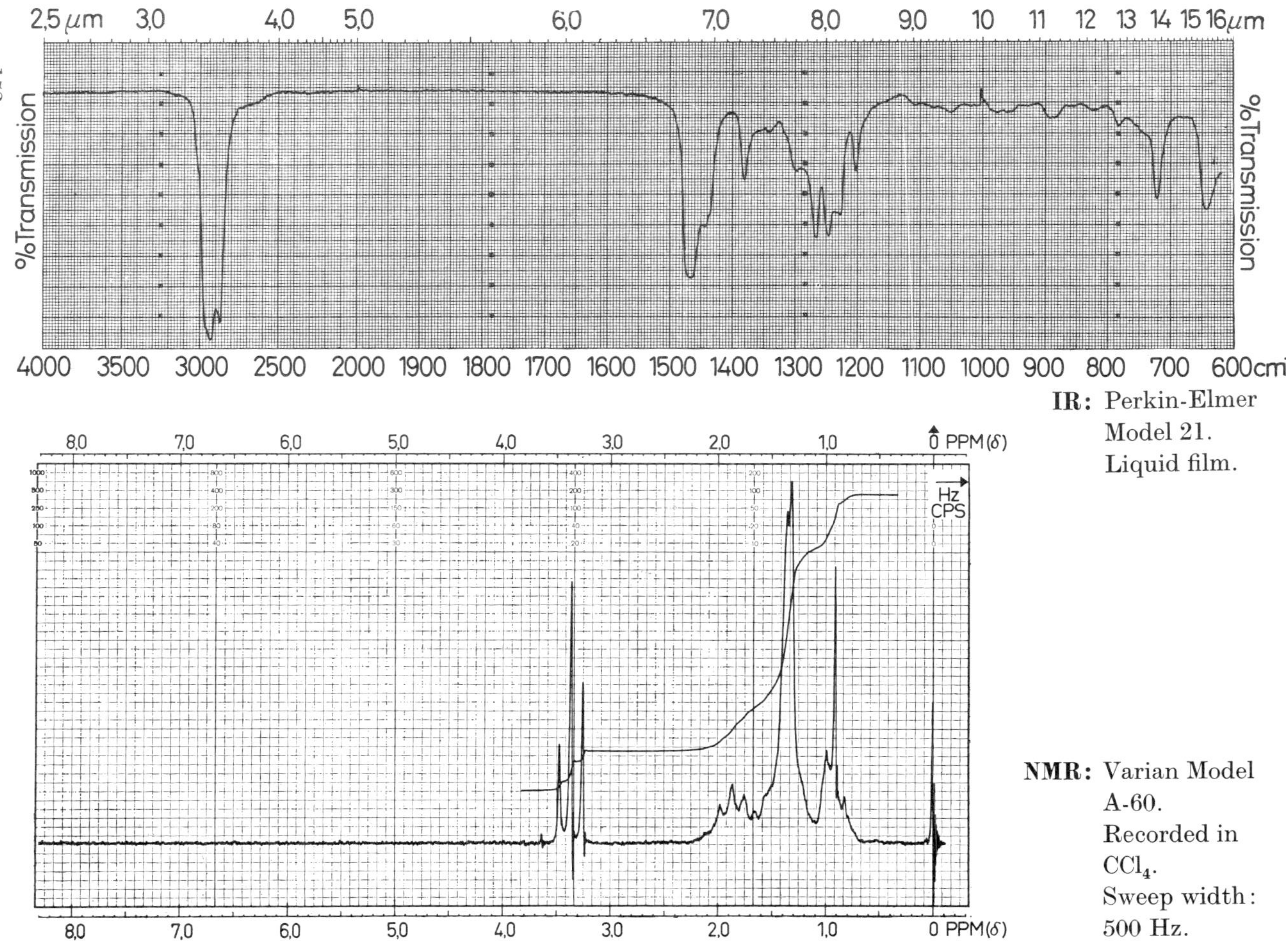

IR: Perkin-Elmer Model 21. Liquid film.

NMR: Varian Model A-60. Recorded in CCl_4. Sweep width: 500 Hz.

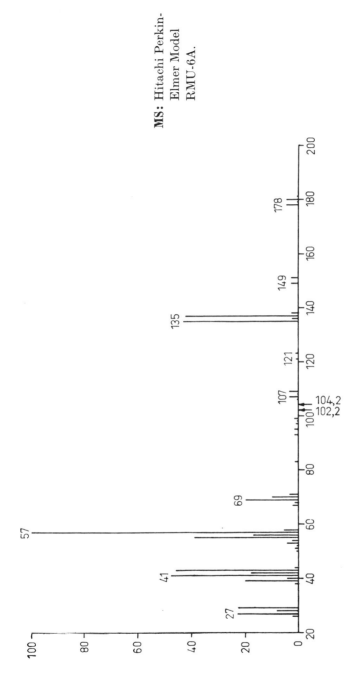

MS: Hitachi Perkin-Elmer Model RMU-6A.

UV: Perkin-Elmer Model 137 UV. Recorded in C_2H_5OH. >210 nm none

Problem 30

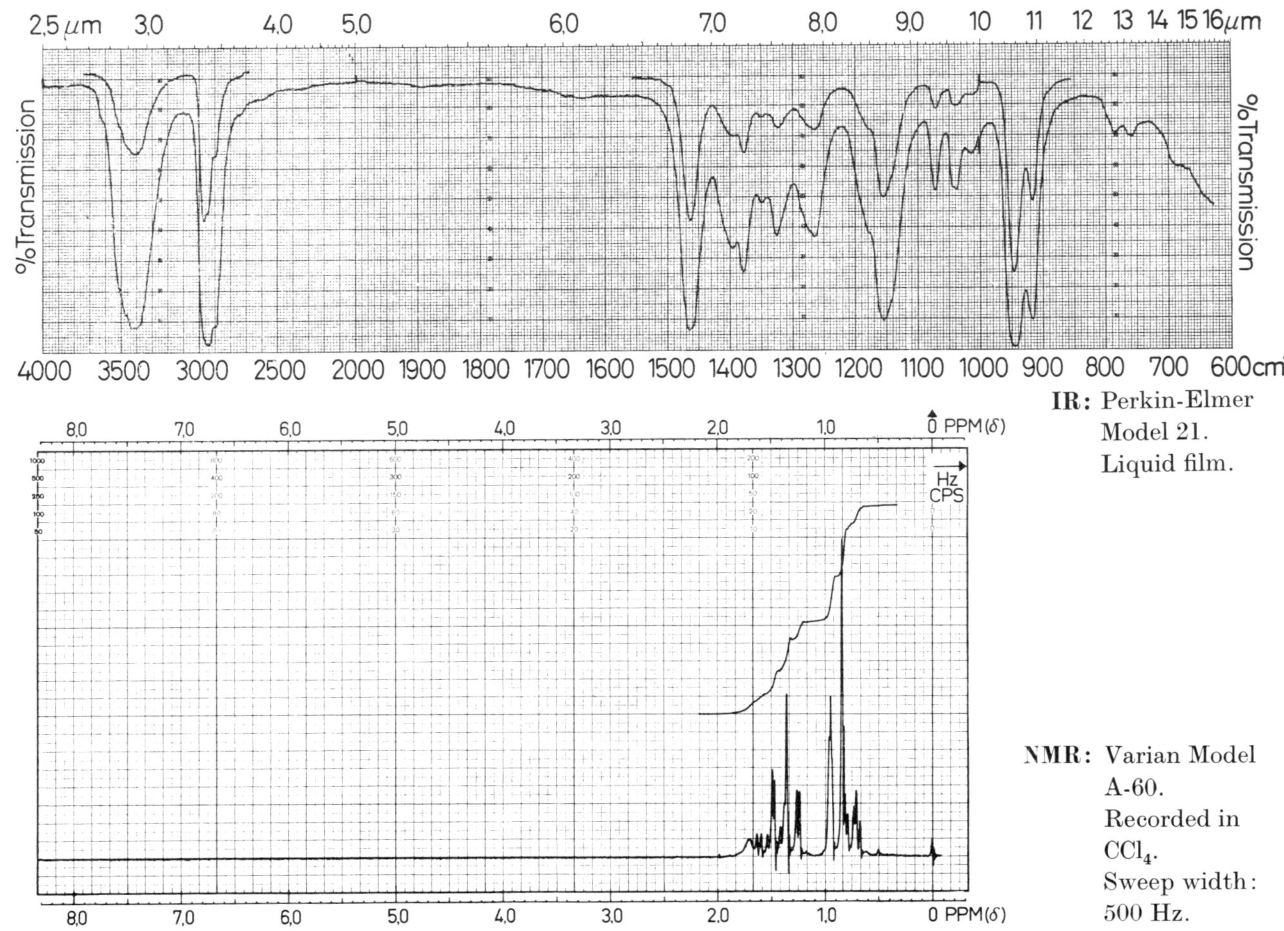

IR: Perkin-Elmer Model 21. Liquid film.

NMR: Varian Model A-60. Recorded in CCl_4. Sweep width: 500 Hz.

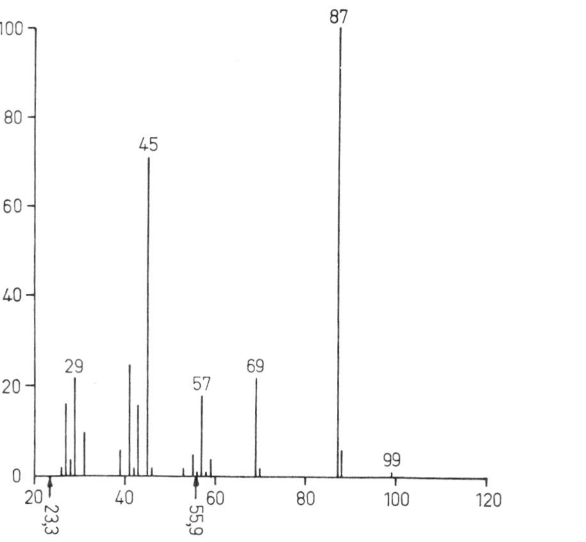

MS: Hitachi Perkin-Elmer Model RMU-6A.

UV: Perkin-Elmer Model 137 UV. Recorded in C_2H_5OH. >210 nm none

Problem 31

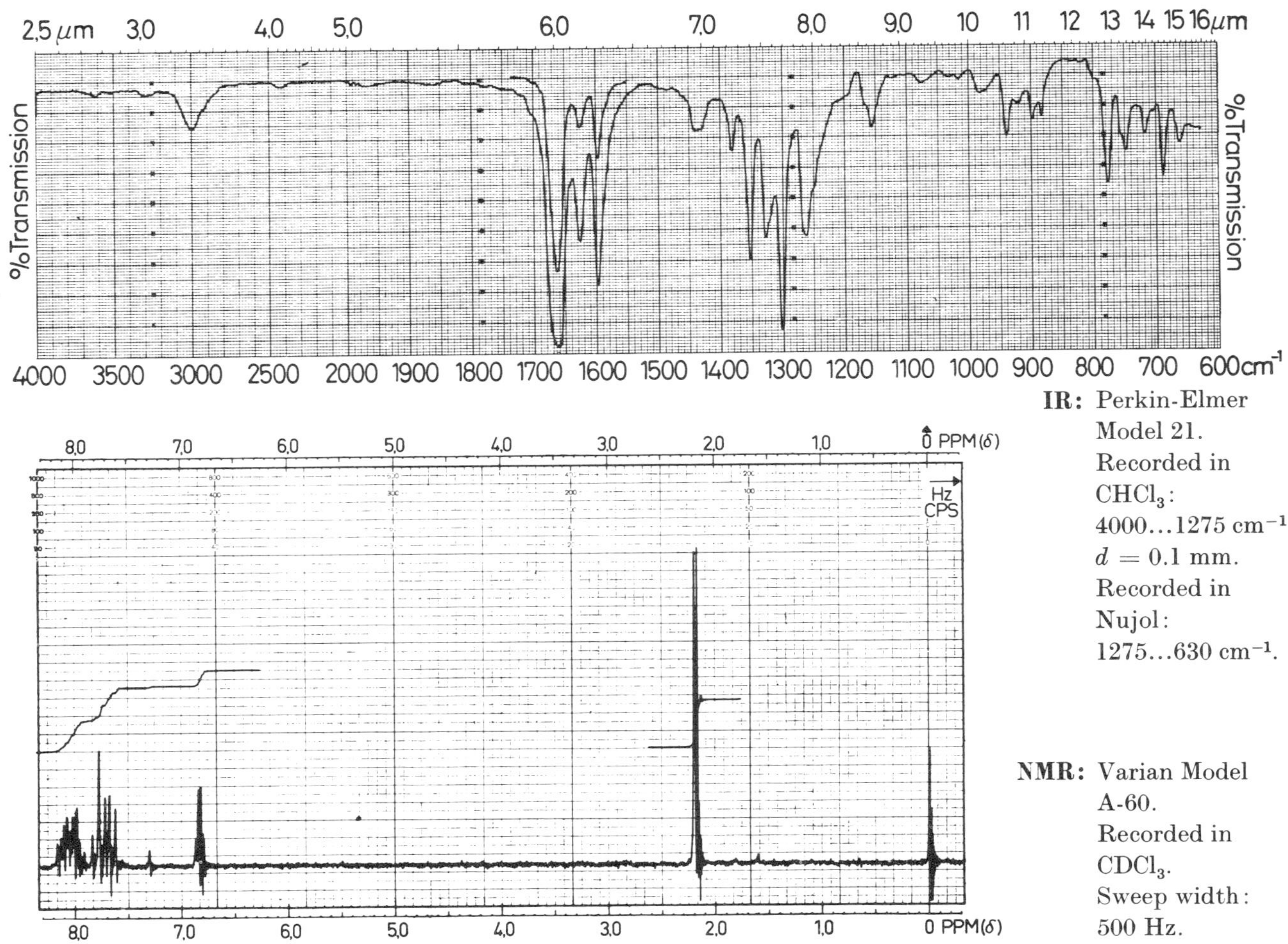

IR: Perkin-Elmer Model 21. Recorded in $CHCl_3$: 4000…1275 cm^{-1} $d = 0.1$ mm. Recorded in Nujol: 1275…630 cm^{-1}.

NMR: Varian Model A-60. Recorded in $CDCl_3$. Sweep width: 500 Hz.

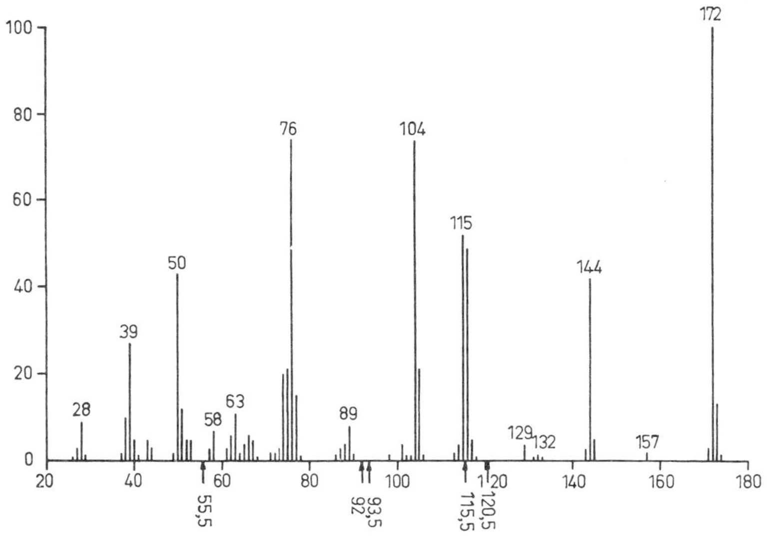

MS: Hitachi Perkin-Elmer Model RMU-6D.

UV: Perkin-Elmer Model 137 UV. Recorded in C_2H_5OH.

λ_{max}	log ε
246 nm	4.2
264 nm s	4.1
335 nm	3.4

Problem 32

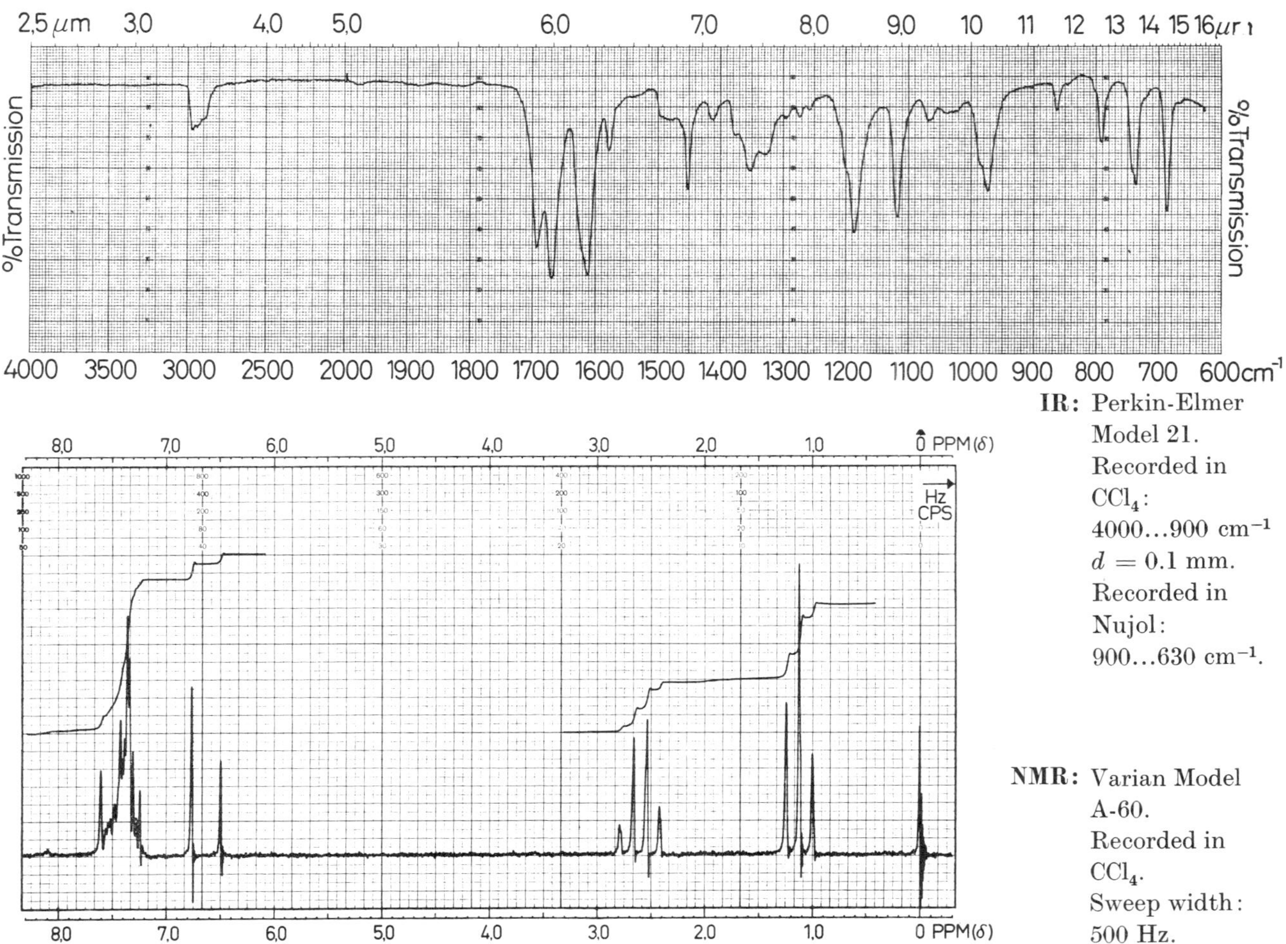

IR: Perkin-Elmer Model 21. Recorded in CCl_4: 4000…900 cm^{-1} $d = 0.1$ mm. Recorded in Nujol: 900…630 cm^{-1}.

NMR: Varian Model A-60. Recorded in CCl_4. Sweep width: 500 Hz.

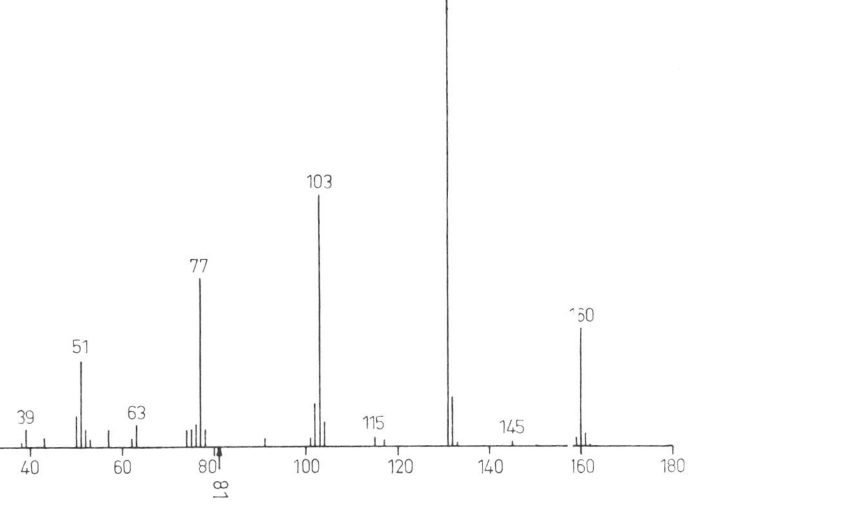

MS: Hitachi Perkin-Elmer Model RMU-6A.

UV: Perkin-Elmer Model 137 UV. Recorded in C_2H_5OH.

λ_{max}	log ε
224 nm	4.1
287 nm	4.3

Problem 33

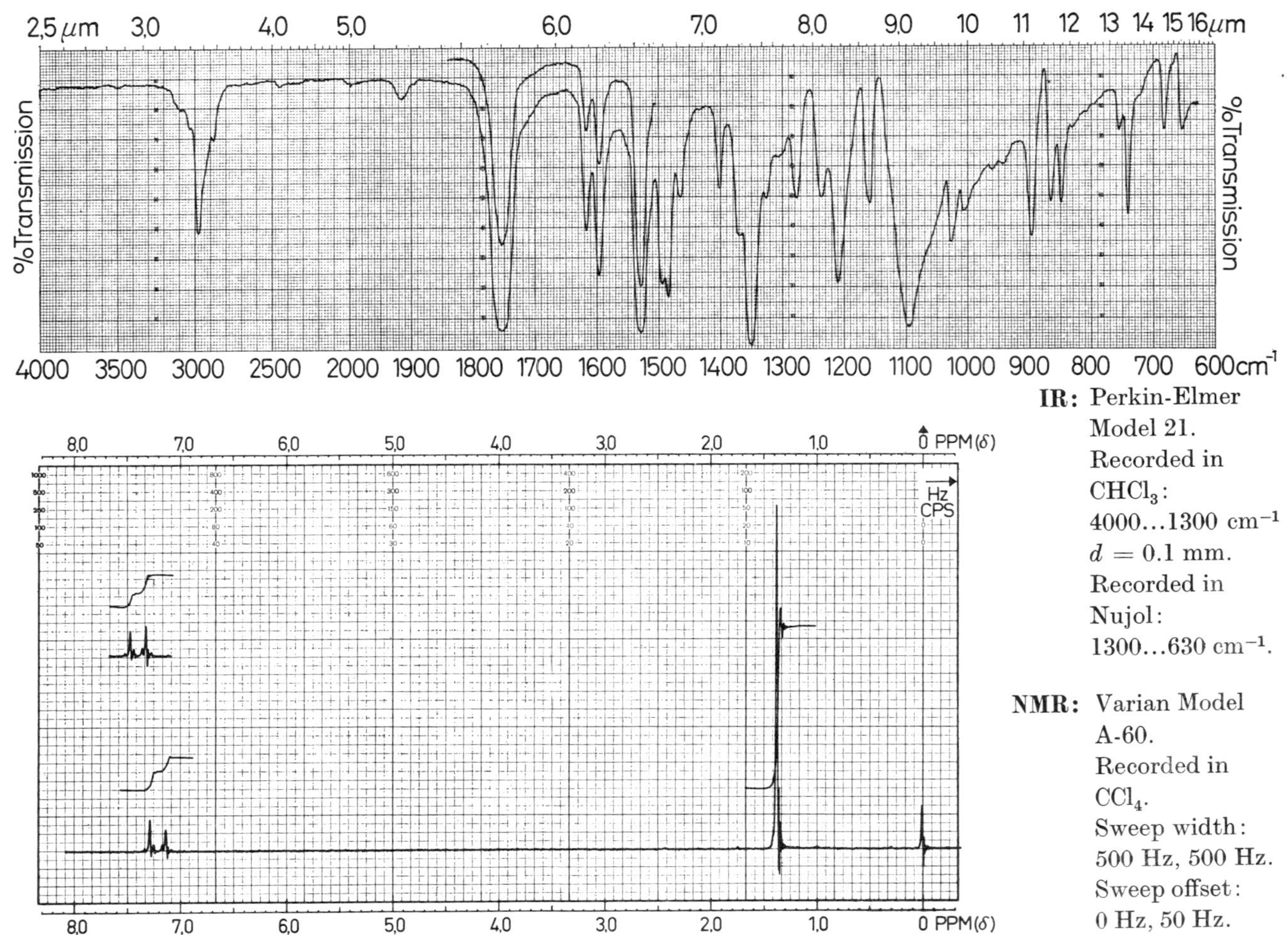

IR: Perkin-Elmer Model 21. Recorded in $CHCl_3$: 4000…1300 cm^{-1} $d = 0.1$ mm. Recorded in Nujol: 1300…630 cm^{-1}.

NMR: Varian Model A-60. Recorded in CCl_4. Sweep width: 500 Hz, 500 Hz. Sweep offset: 0 Hz, 50 Hz.

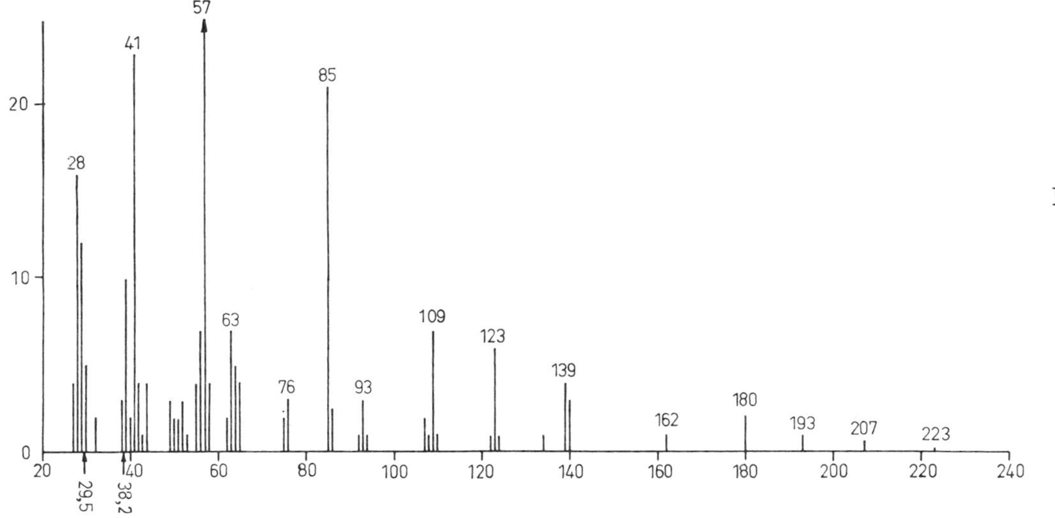

MS: Hitachi Perkin-Elmer Model RMU-6A.

UV: Perkin-Elmer Model 137 UV. Recorded in C_2H_5OH.

λ_{max}	$\log \varepsilon$
269 nm	4.0

Problem 34

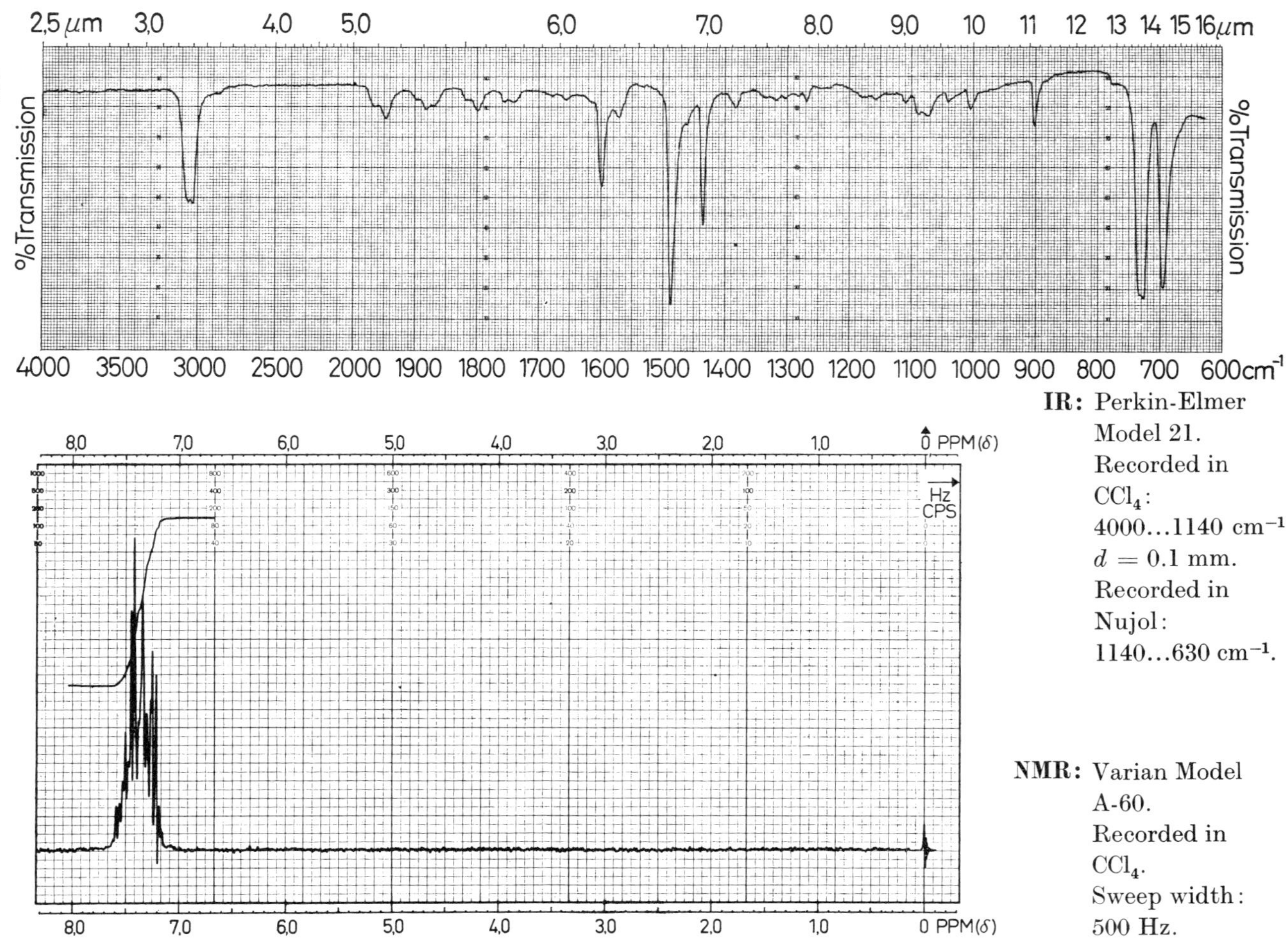

IR: Perkin-Elmer Model 21. Recorded in CCl_4: 4000…1140 cm^{-1} $d = 0.1$ mm. Recorded in Nujol: 1140…630 cm^{-1}.

NMR: Varian Model A-60. Recorded in CCl_4. Sweep width: 500 Hz.

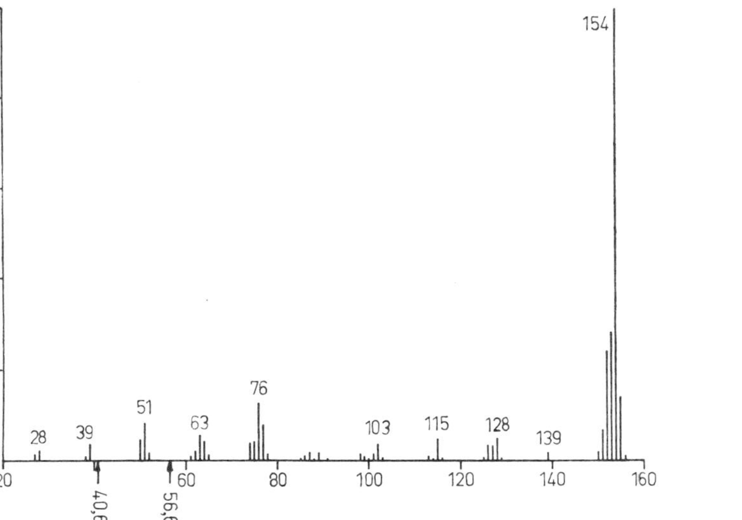

MS: Hitachi Perkin-Elmer Model RMU-6A.

UV: Perkin-Elmer Model 137 UV. Recorded in C_2H_5OH.

λ_{max}	$\log \varepsilon$
250 nm	4.2

Problem 35

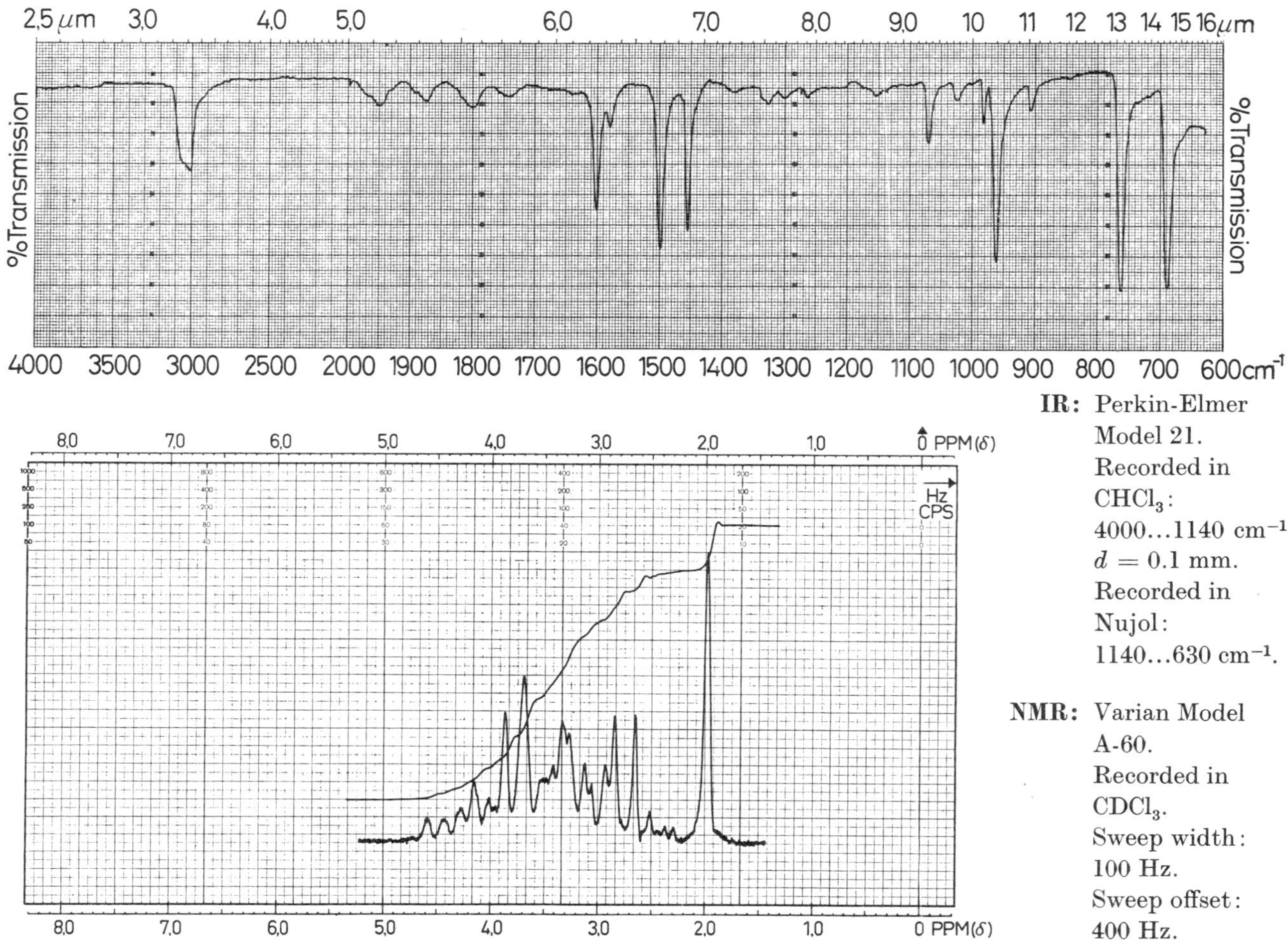

IR: Perkin-Elmer Model 21. Recorded in $CHCl_3$: 4000...1140 cm^{-1} $d = 0.1$ mm. Recorded in Nujol: 1140...630 cm^{-1}.

NMR: Varian Model A-60. Recorded in $CDCl_3$. Sweep width: 100 Hz. Sweep offset: 400 Hz.

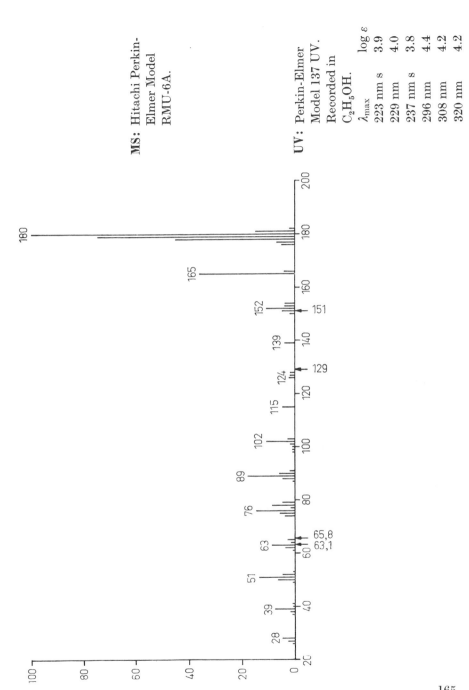

MS: Hitachi Perkin-Elmer Model RMU-6A.

UV: Perkin-Elmer Model 137 UV. Recorded in C_2H_5OH.

λ_{max}	log ε
223 nm s	3.9
229 nm	4.0
237 nm s	3.8
296 nm	4.4
308 nm	4.2
320 nm	4.2

Problem 36

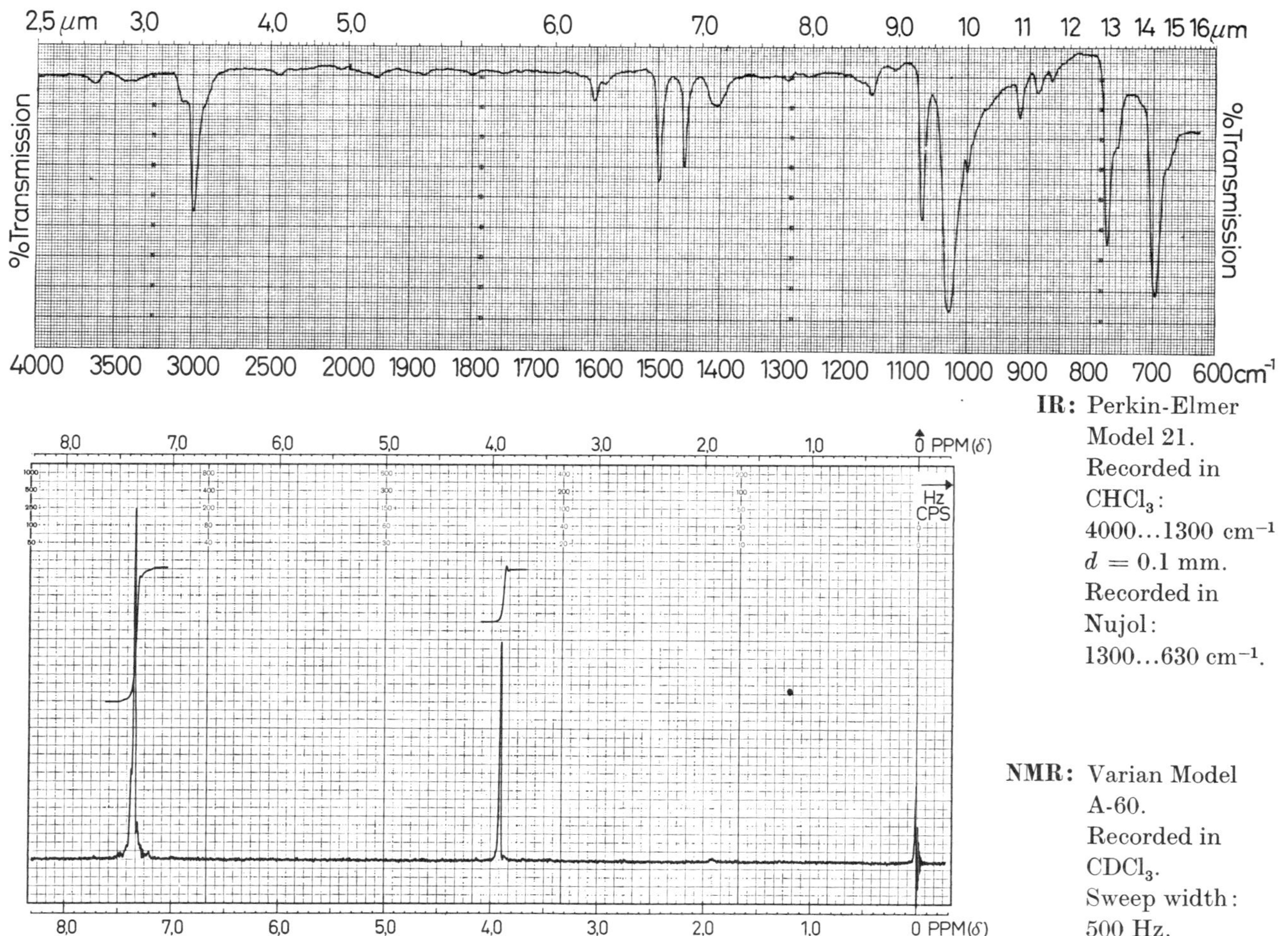

IR: Perkin-Elmer Model 21. Recorded in $CHCl_3$: 4000…1300 cm^{-1} $d = 0.1$ mm. Recorded in Nujol: 1300…630 cm^{-1}.

NMR: Varian Model A-60. Recorded in $CDCl_3$. Sweep width: 500 Hz.

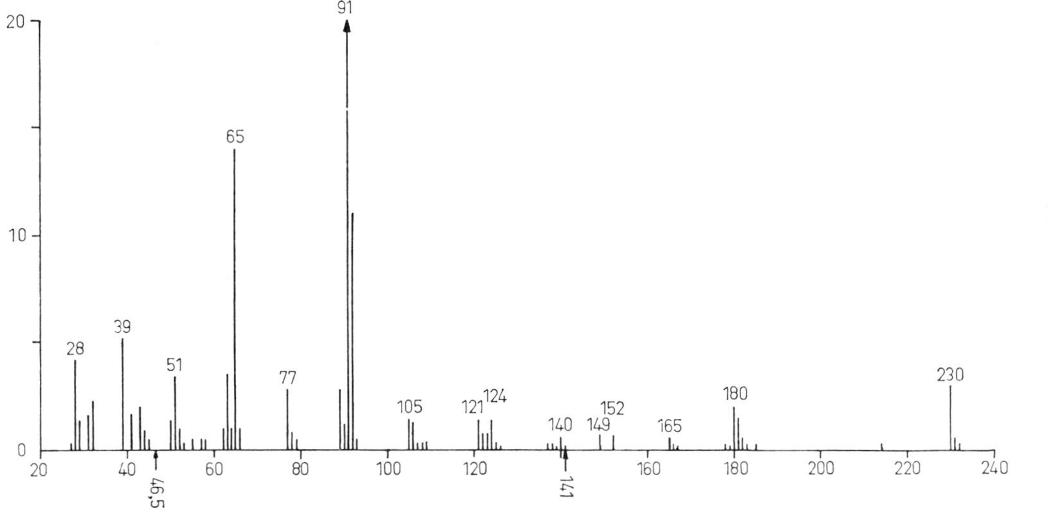

MS: Hitachi Perkin-Elmer Model RMU-6D.

UV: Perkin-Elmer Model 137 UV. Recorded in C_2H_5OH.

λ_{max}	$\log \varepsilon$
223 nm	4.2
253 nm s	3.8
260 nm	3.8
266 nm	3.7
270 nm	3.5

Problem 37

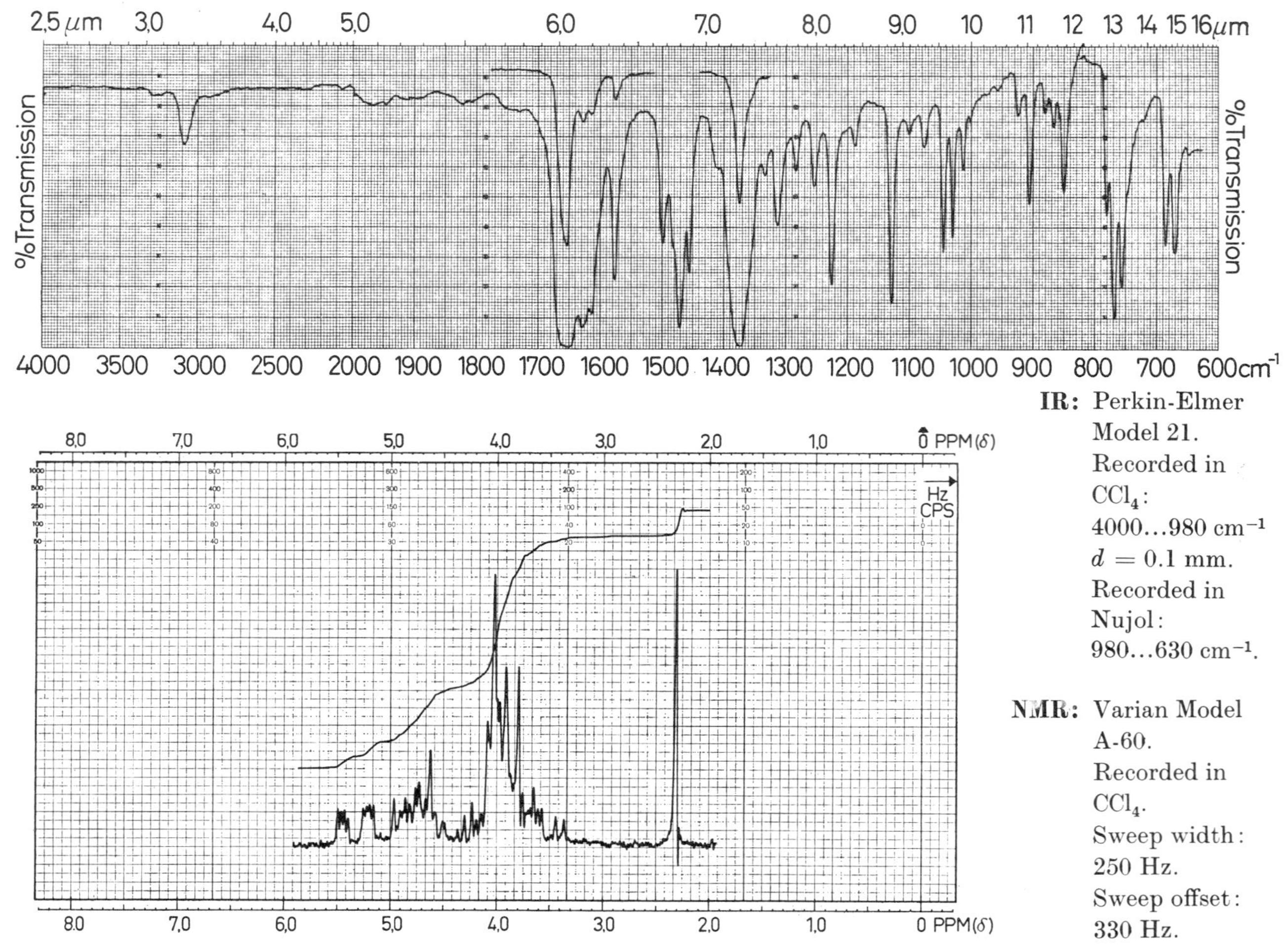

IR: Perkin-Elmer Model 21. Recorded in CCl_4: 4000…980 cm^{-1} $d = 0.1$ mm. Recorded in Nujol: 980…630 cm^{-1}.

NMR: Varian Model A-60. Recorded in CCl_4. Sweep width: 250 Hz. Sweep offset: 330 Hz.

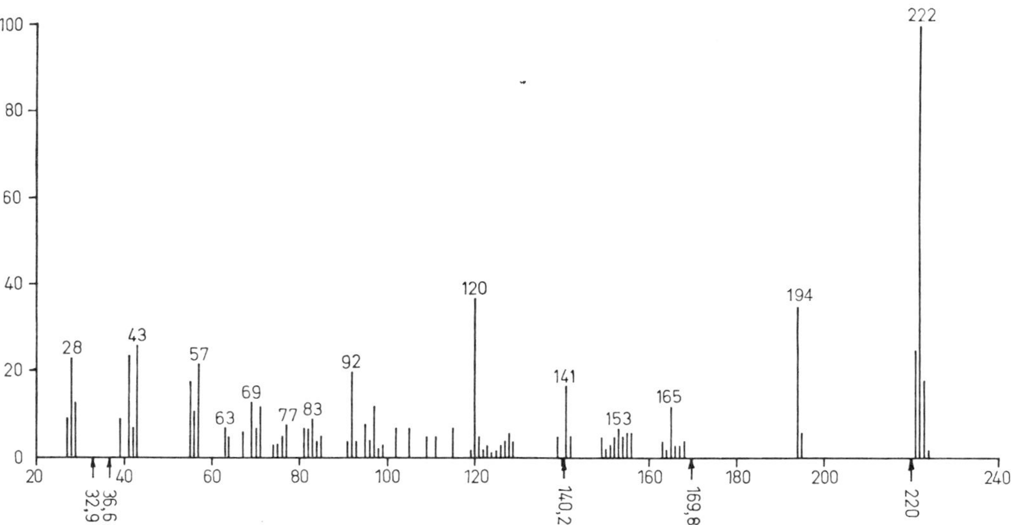

MS: Hitachi Perkin-Elmer Model RMU-6A.

UV: Perkin-Elmer Model 137 UV. Recorded in C_2H_5OH.

λ_{max}	log ε
214 nm s	3.5
250 nm s	3.8
295 nm	3.9
309 nm	3.8

Problem 38

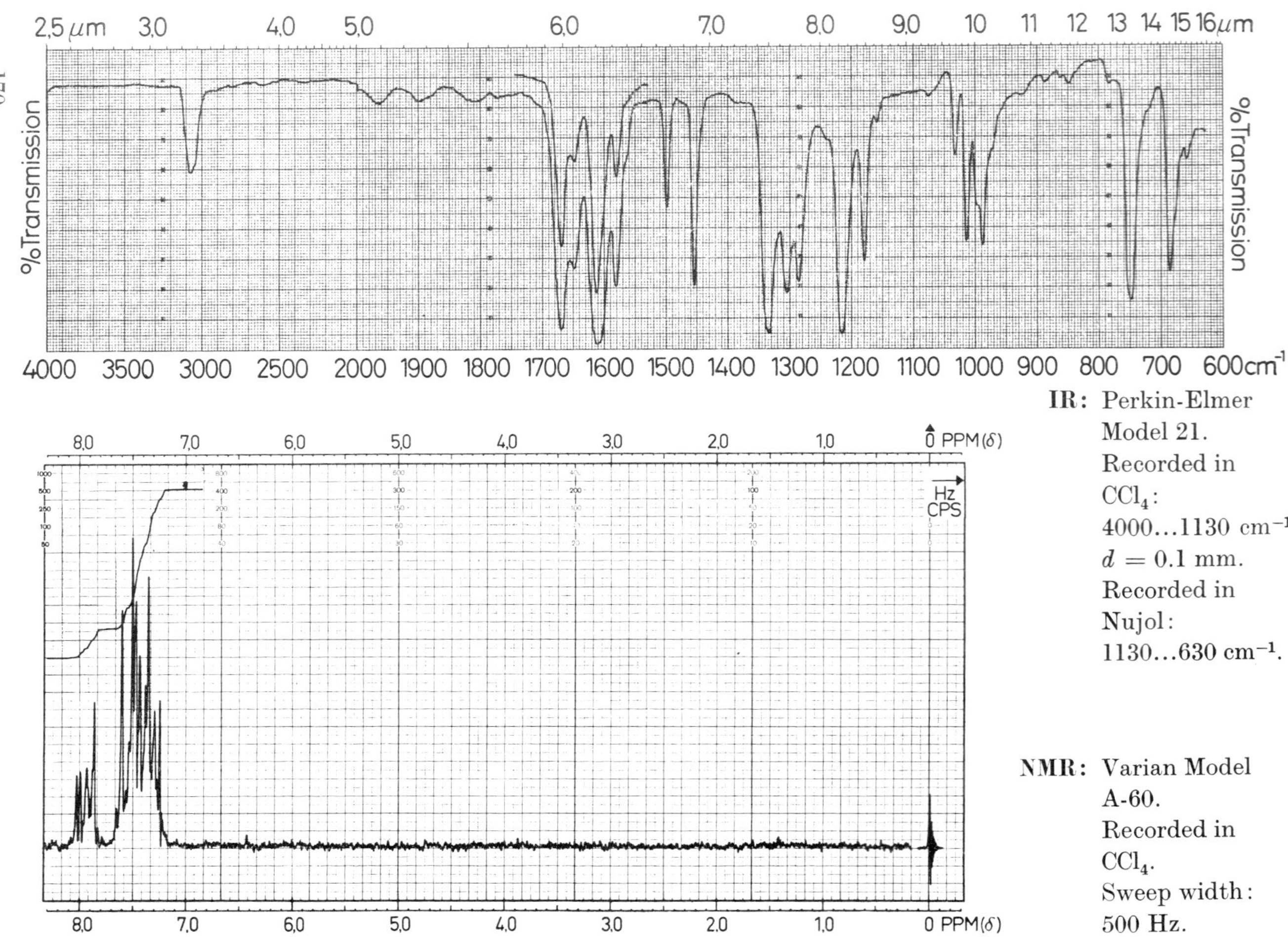

IR: Perkin-Elmer Model 21. Recorded in CCl_4: 4000...1130 cm^{-1} $d = 0.1$ mm. Recorded in Nujol: 1130...630 cm^{-1}.

NMR: Varian Model A-60. Recorded in CCl_4. Sweep width: 500 Hz.

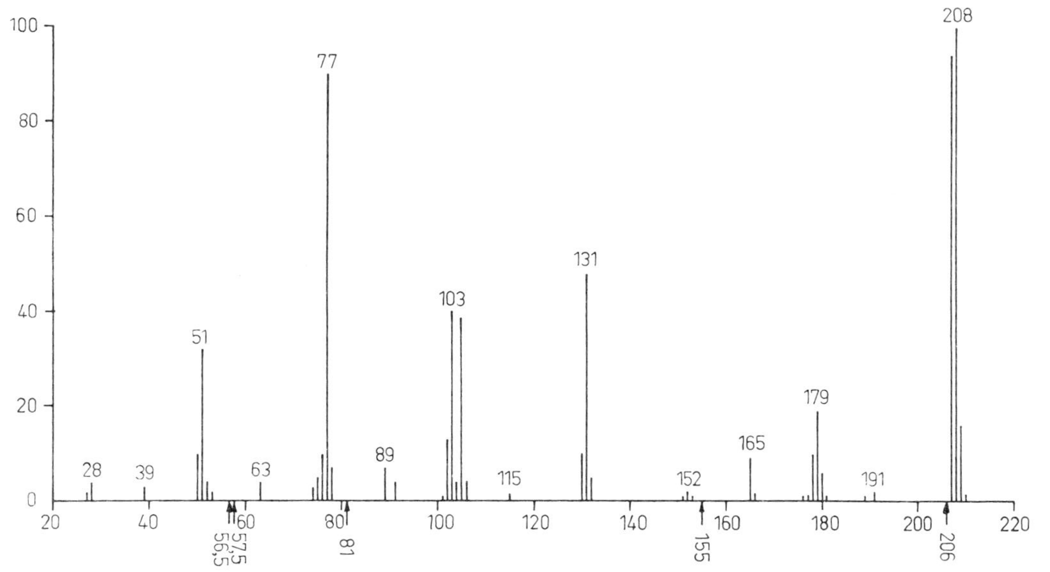

MS: Hitachi Perkin-Elmer Model RMU-6D.

UV: Perkin-Elmer Model 137 UV. Recorded in C_2H_5OH.

λ_{max}	$\log \varepsilon$
228 nm	3.9
310 nm	4.3

Problem 39

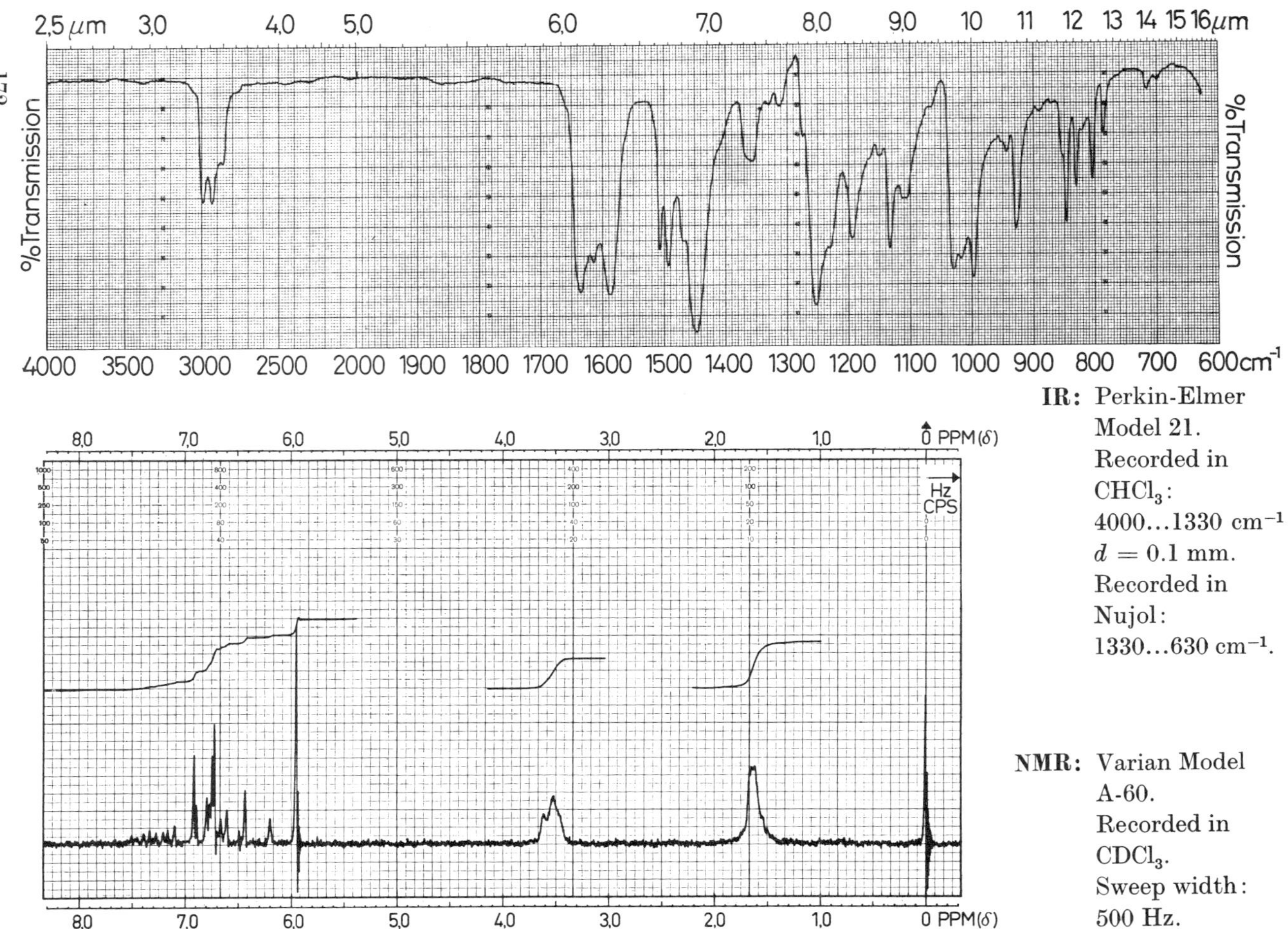

IR: Perkin-Elmer Model 21. Recorded in $CHCl_3$: 4000…1330 cm^{-1} $d = 0.1$ mm. Recorded in Nujol: 1330…630 cm^{-1}.

NMR: Varian Model A-60. Recorded in $CDCl_3$. Sweep width: 500 Hz.

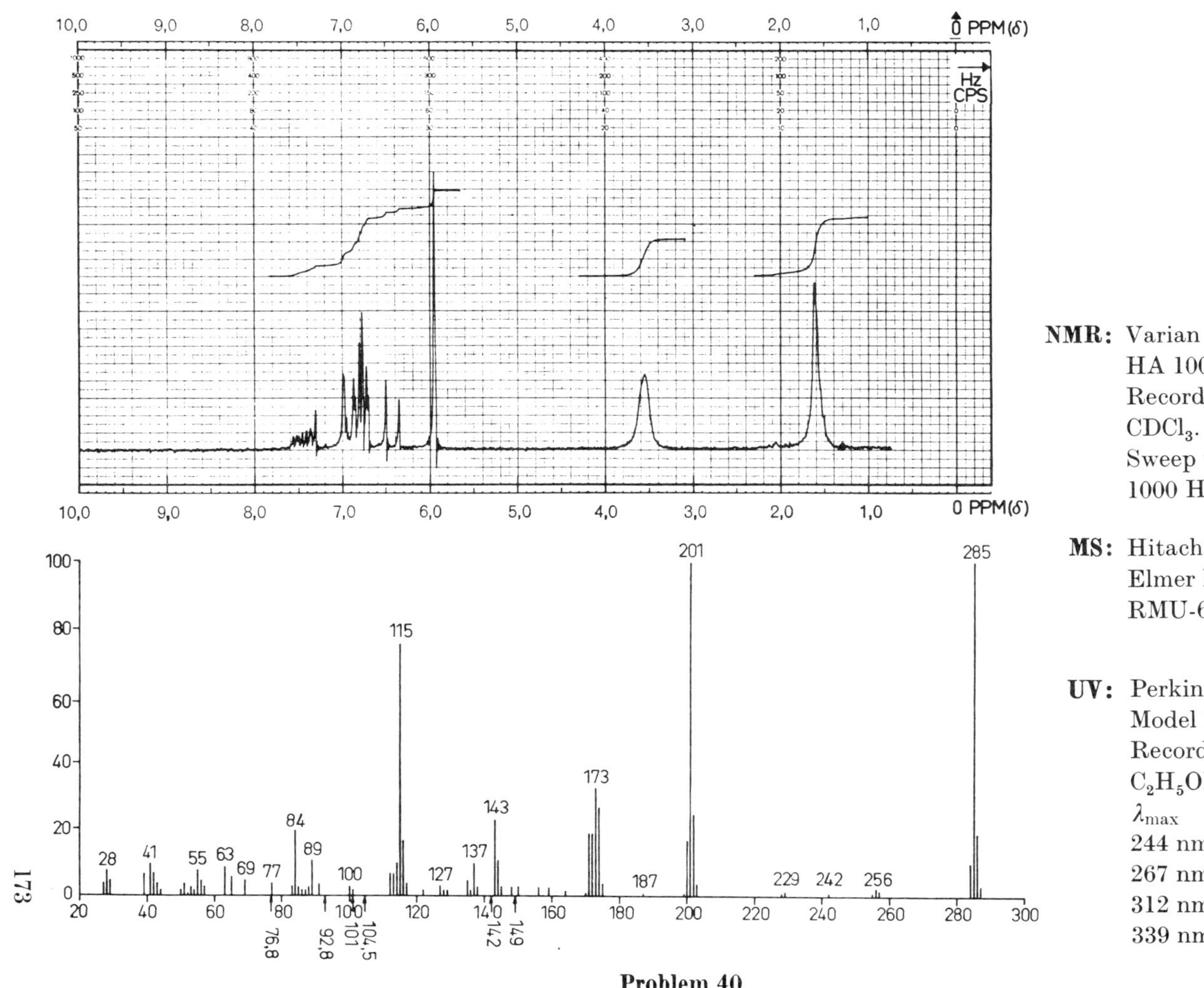

NMR: Varian Model HA 100. Recorded in $CDCl_3$. Sweep width: 1000 Hz.

MS: Hitachi Perkin-Elmer Model RMU-6A.

UV: Perkin-Elmer Model 137 UV. Recorded in C_2H_5OH.

λ_{max}	log ε
244 nm s	3.5
267 nm	3.6
312 nm	3.8
339 nm	3.9

Problem 40

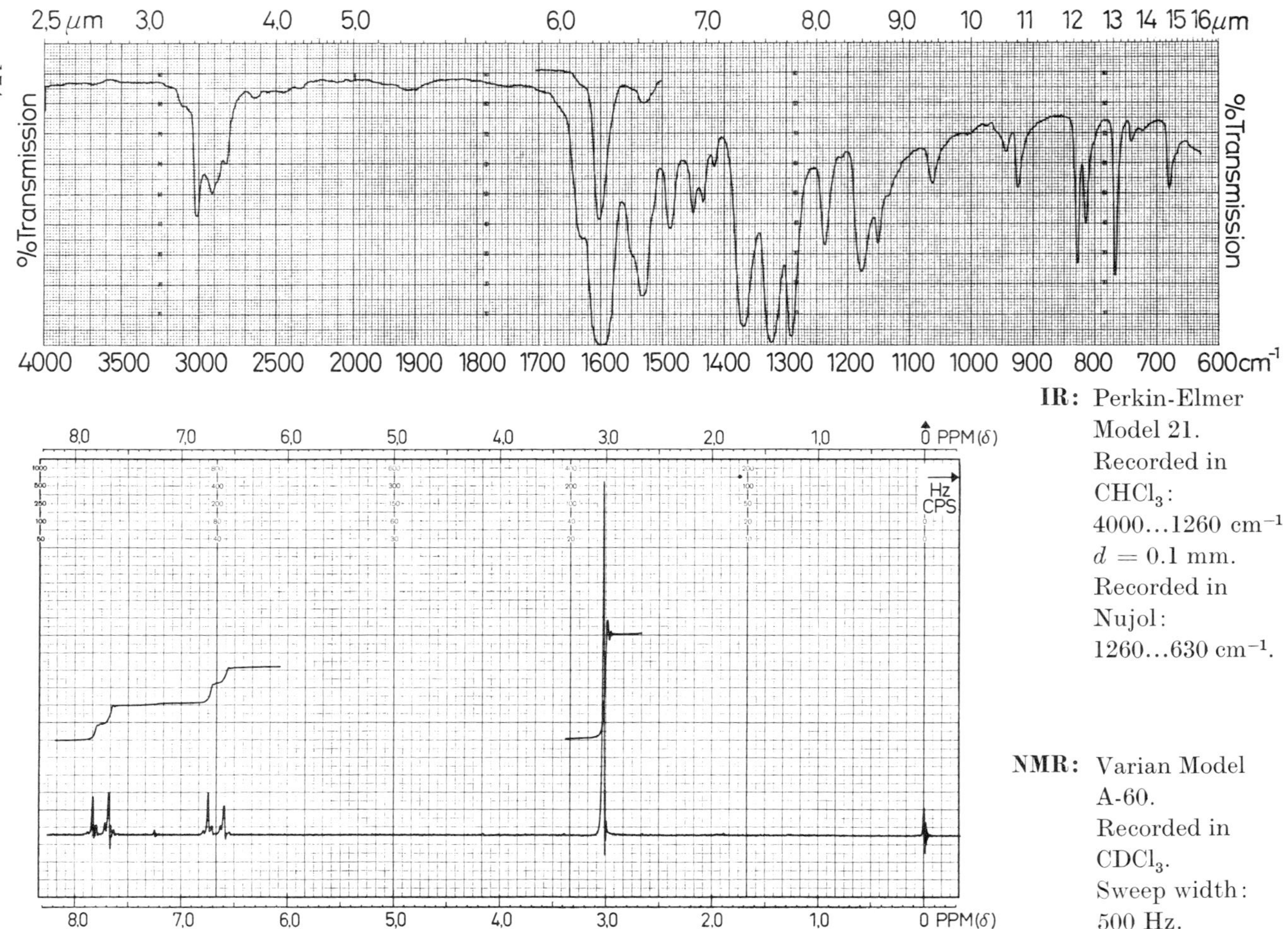

IR: Perkin-Elmer Model 21. Recorded in $CHCl_3$: 4000...1260 cm^{-1} $d = 0.1$ mm. Recorded in Nujol: 1260...630 cm^{-1}.

NMR: Varian Model A-60. Recorded in $CDCl_3$. Sweep width: 500 Hz.

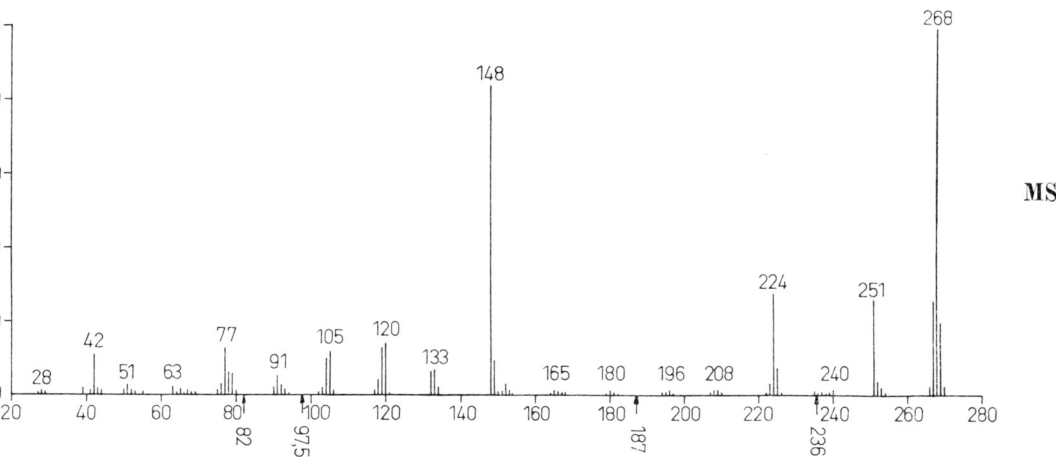

MS: Hitachi Perkin-Elmer Model RMU-6A.

UV: Perkin-Elmer Model 137 UV. Recorded in C_2H_5OH.

λ_{max}	log ε
244 nm	4.2
310 nm s	4.0
367 nm	4.5

Problem 41

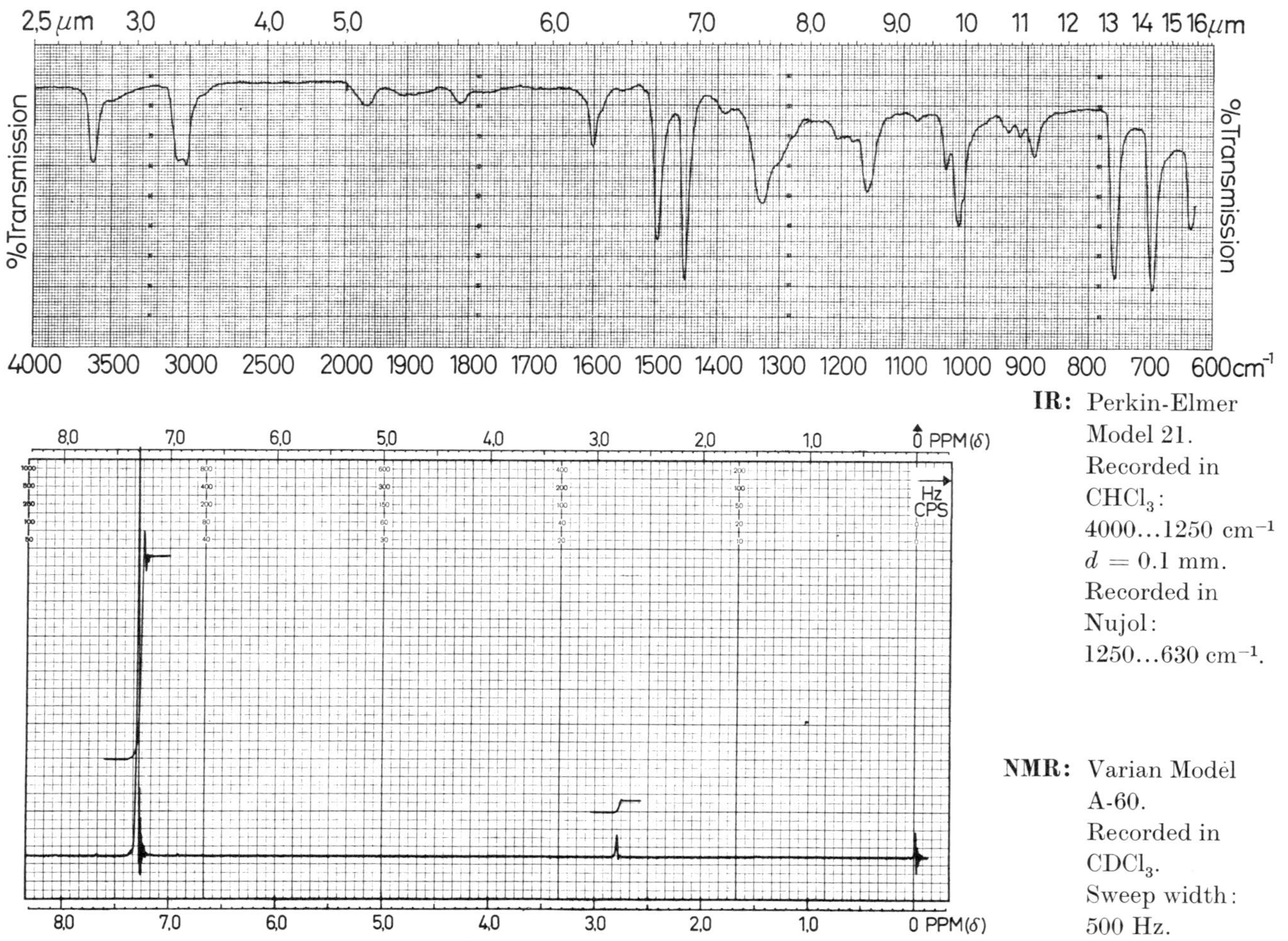

IR: Perkin-Elmer Model 21. Recorded in $CHCl_3$: 4000...1250 cm^{-1} $d = 0.1$ mm. Recorded in Nujol: 1250...630 cm^{-1}.

NMR: Varian Model A-60. Recorded in $CDCl_3$. Sweep width: 500 Hz.

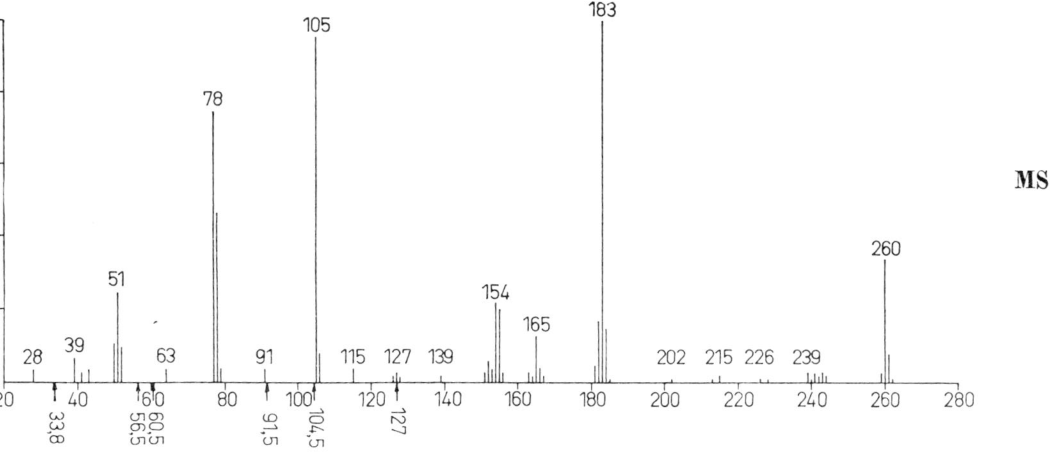

MS: Hitachi Perkin-Elmer Model RMU-6A.

UV: Perkin-Elmer Model 137 UV. Recorded in C_2H_5OH.

λ_{max}	$\log \varepsilon$
214 nm	4.4
248 nm	3.6
252 nm	3.7
258 nm	3.8
265 nm	3.7
268 nm	3.5

Problem 42

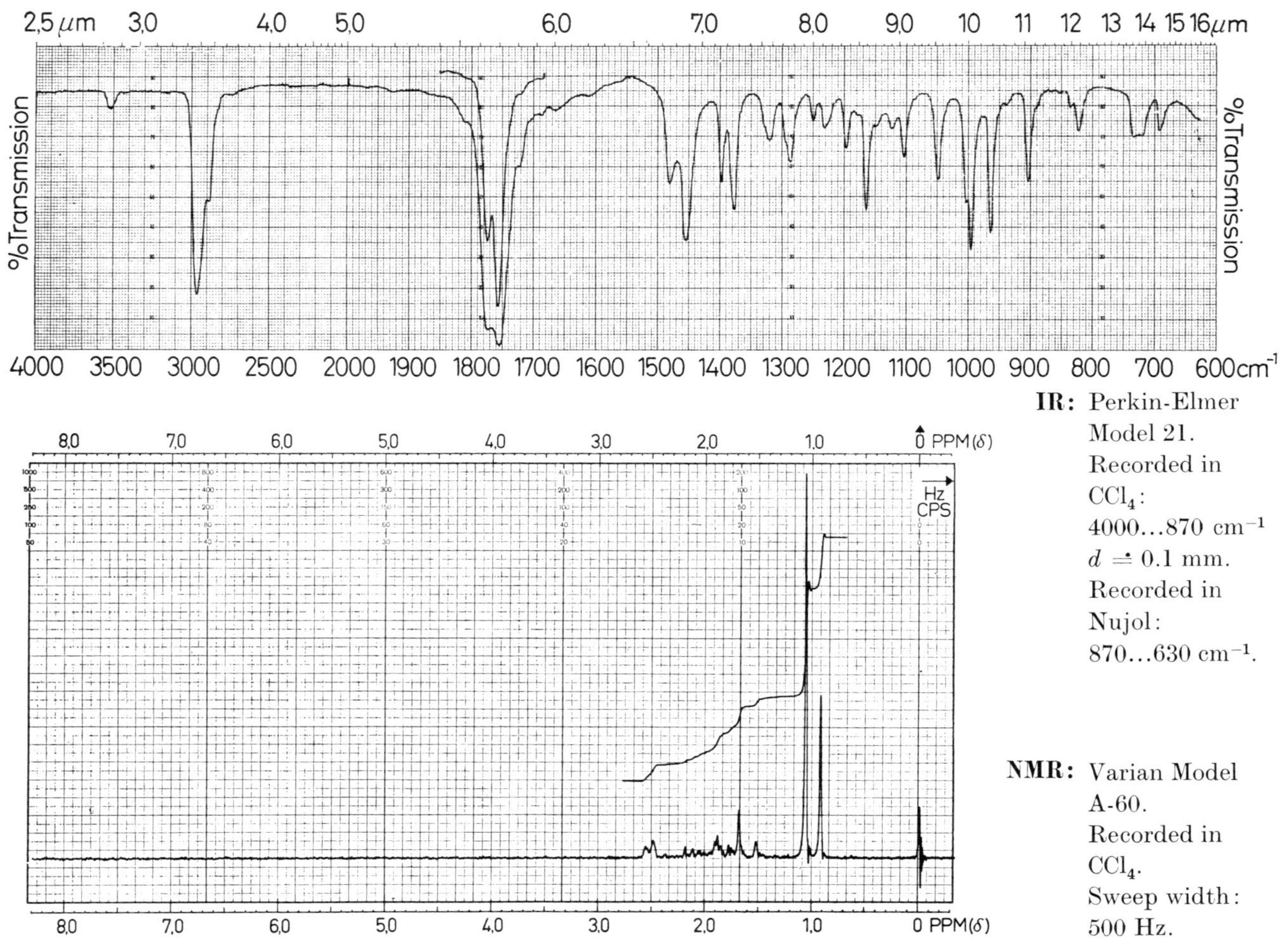

IR: Perkin-Elmer Model 21. Recorded in CCl_4: 4000...870 cm^{-1} $d \doteq 0.1$ mm. Recorded in Nujol: 870...630 cm^{-1}.

NMR: Varian Model A-60. Recorded in CCl_4. Sweep width: 500 Hz.

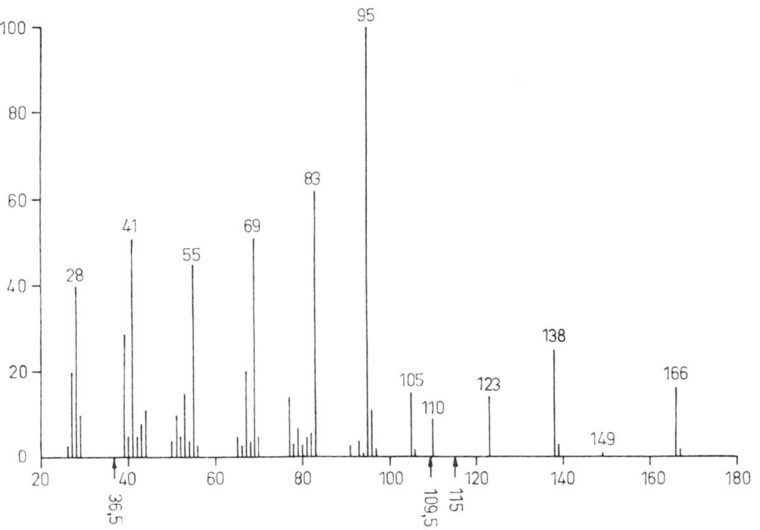

MS: Hitachi Perkin-Elmer Model RMU-6A.

UV: Perkin-Elmer Model 137 UV. Recorded in C_2H_5OH.

λ_{max}	$\log \varepsilon$
269 nm	1.4
279 nm	1.4
457 nm s	1.4
468 nm	1.5

Problem 43

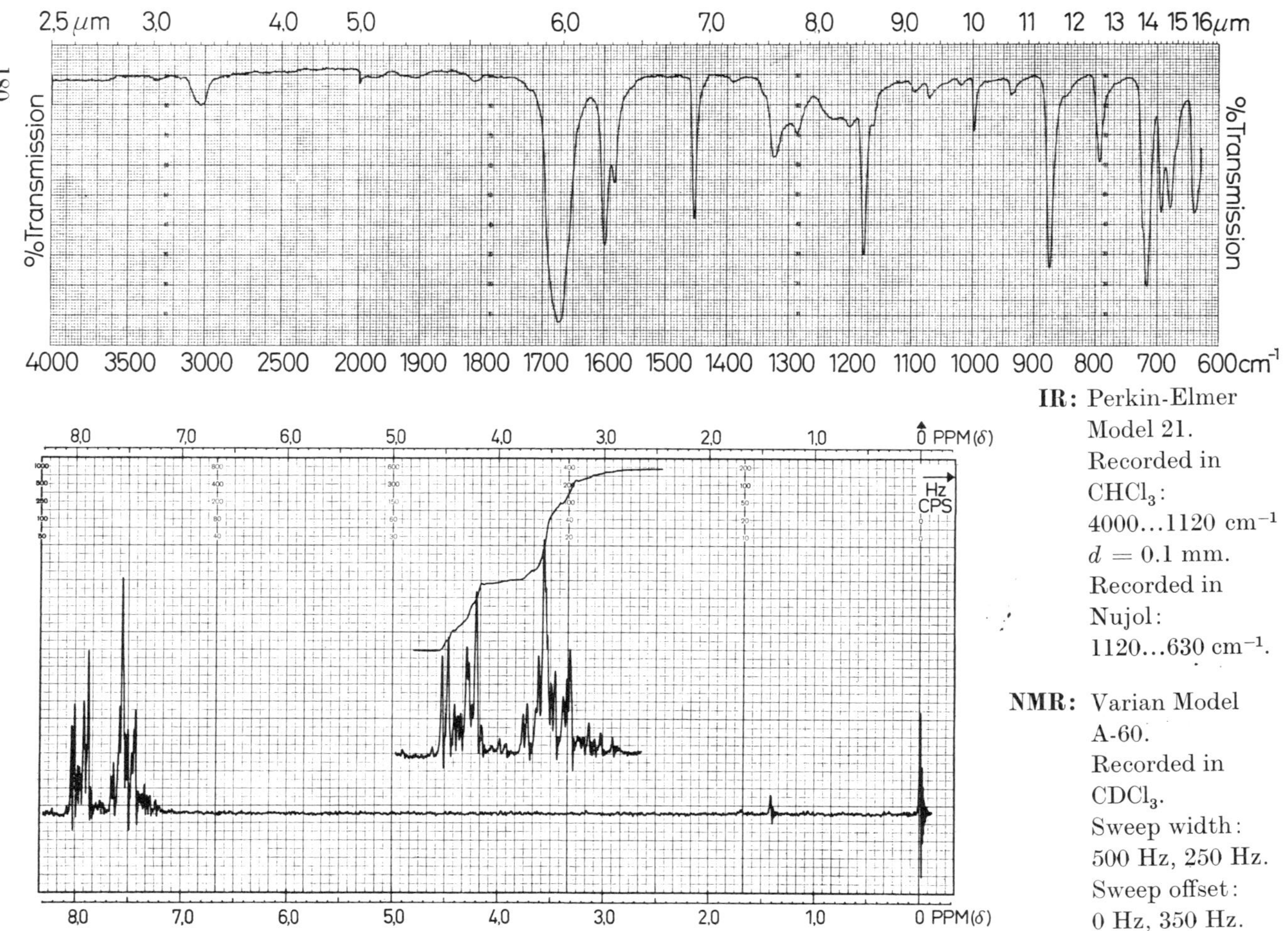

IR: Perkin-Elmer Model 21. Recorded in $CHCl_3$: 4000...1120 cm^{-1} $d = 0.1$ mm. Recorded in Nujol: 1120...630 cm^{-1}.

NMR: Varian Model A-60. Recorded in $CDCl_3$. Sweep width: 500 Hz, 250 Hz. Sweep offset: 0 Hz, 350 Hz.

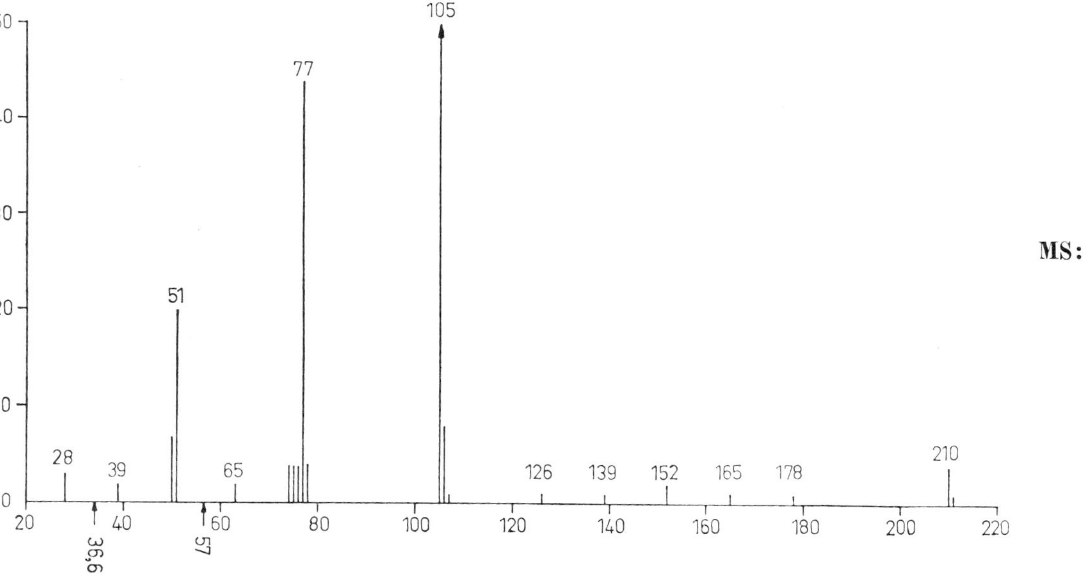

MS: Hitachi Perkin-Elmer Model RMU-6A.

UV: Perkin-Elmer Model 137 UV. Recorded in C_2H_5OH.

λ_{max}	$\log \varepsilon$
259 nm	4.1

Problem 44

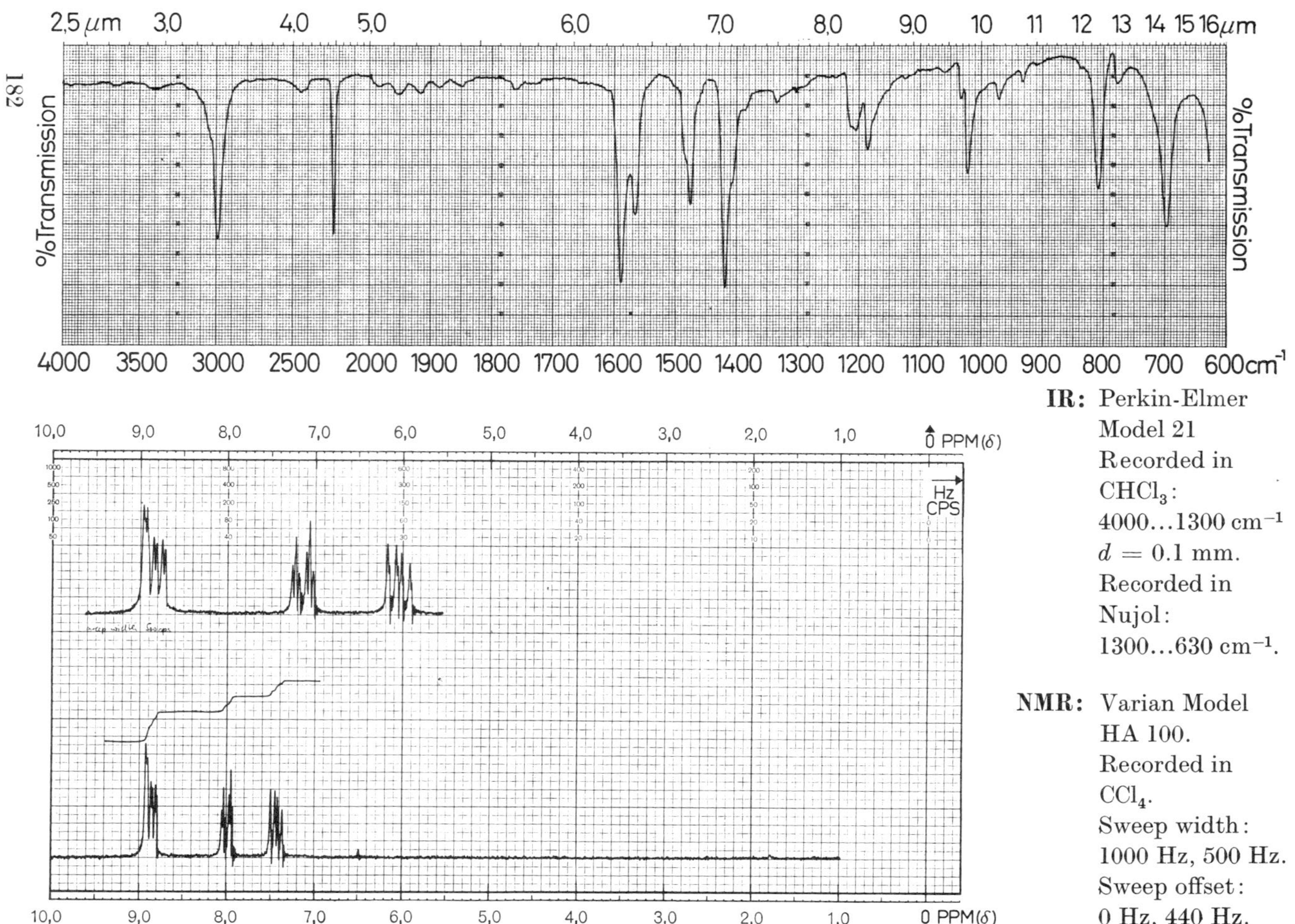

IR: Perkin-Elmer Model 21
Recorded in $CHCl_3$: 4000...1300 cm^{-1}
$d = 0.1$ mm.
Recorded in Nujol: 1300...630 cm^{-1}.

NMR: Varian Model HA 100.
Recorded in CCl_4.
Sweep width: 1000 Hz, 500 Hz.
Sweep offset: 0 Hz, 440 Hz.

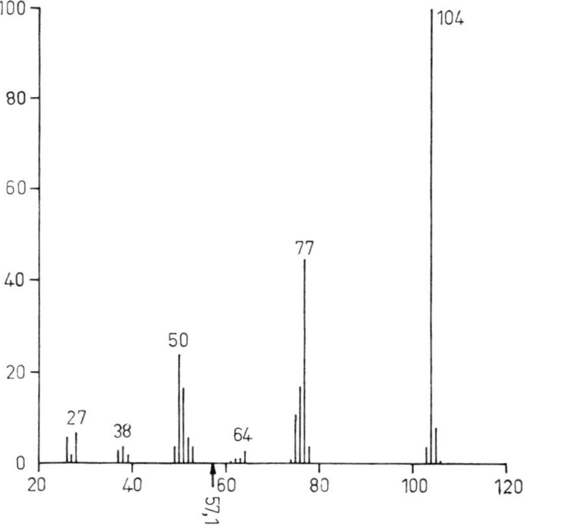

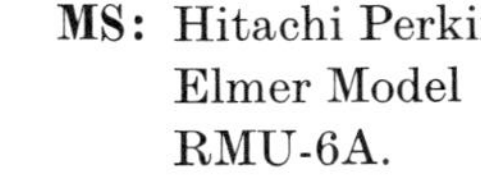

MS: Hitachi Perkin-Elmer Model RMU-6A.

UV: Perkin-Elmer Model 137 UV. Recorded in C_2H_5OH.

λ_{max}	$\log \varepsilon$
218 nm	3.9
226 nm	3.8
259 nm	3.3
265 nm	3.4
272 nm	3.3

Problem 45

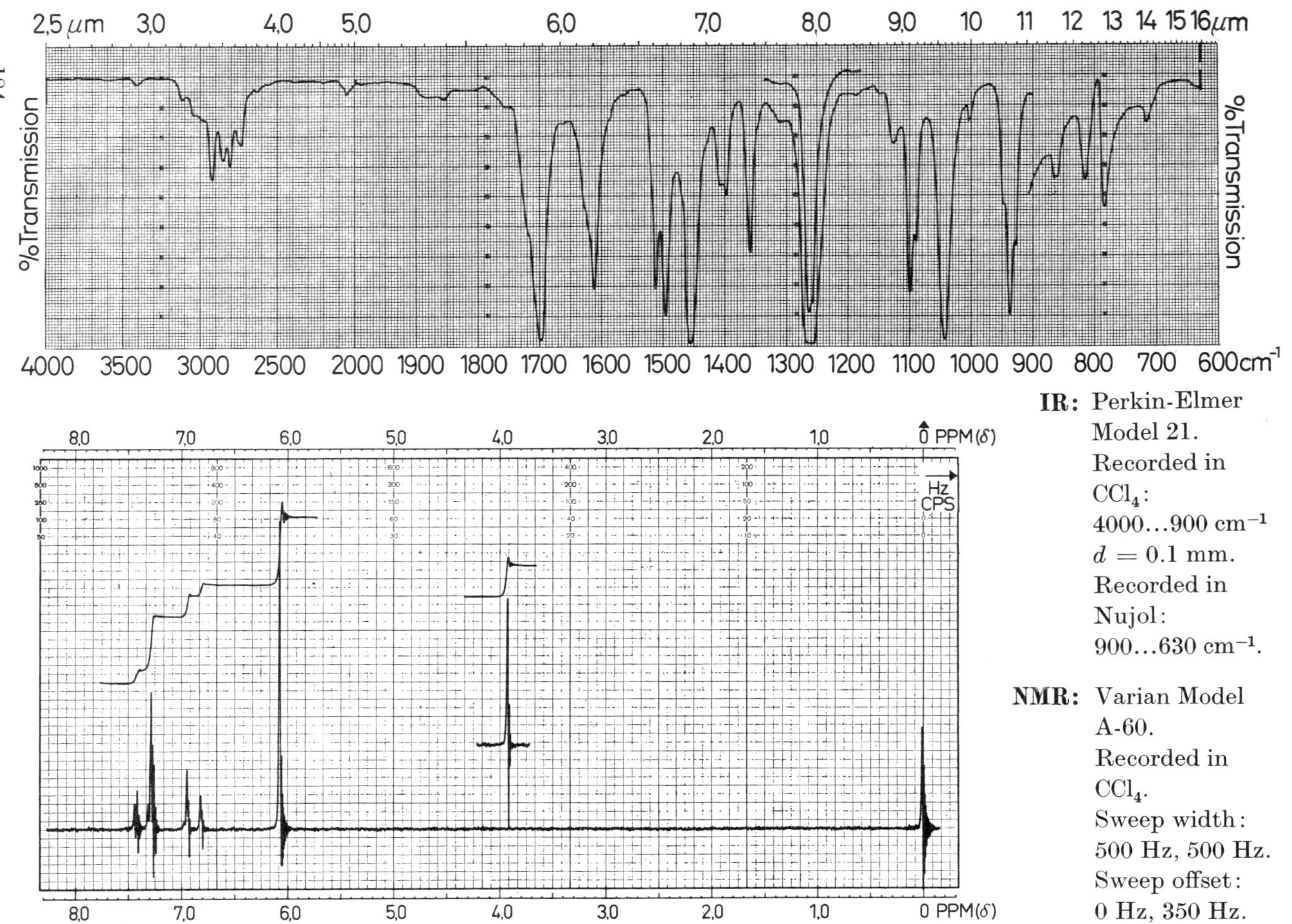

IR: Perkin-Elmer Model 21. Recorded in CCl_4: 4000...900 cm^{-1} $d = 0.1$ mm. Recorded in Nujol: 900...630 cm^{-1}.

NMR: Varian Model A-60. Recorded in CCl_4. Sweep width: 500 Hz, 500 Hz. Sweep offset: 0 Hz, 350 Hz.

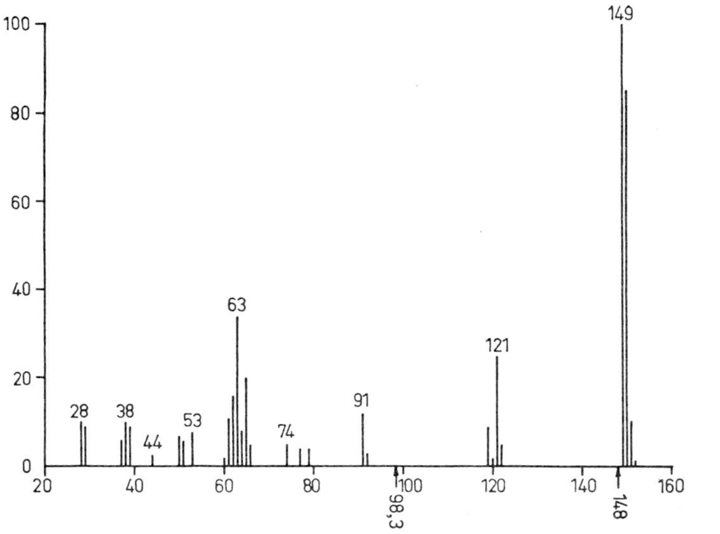

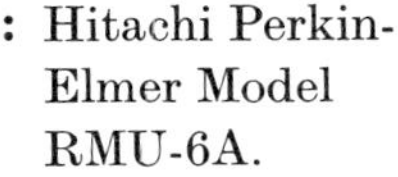

MS: Hitachi Perkin-Elmer Model RMU-6A.

UV: Perkin-Elmer Model 137 UV. Recorded in C_2H_5OH.

λ_{max}	$\log \varepsilon$
230 nm	4.2
274 nm	3.9
315 nm	4.0

Problem 46

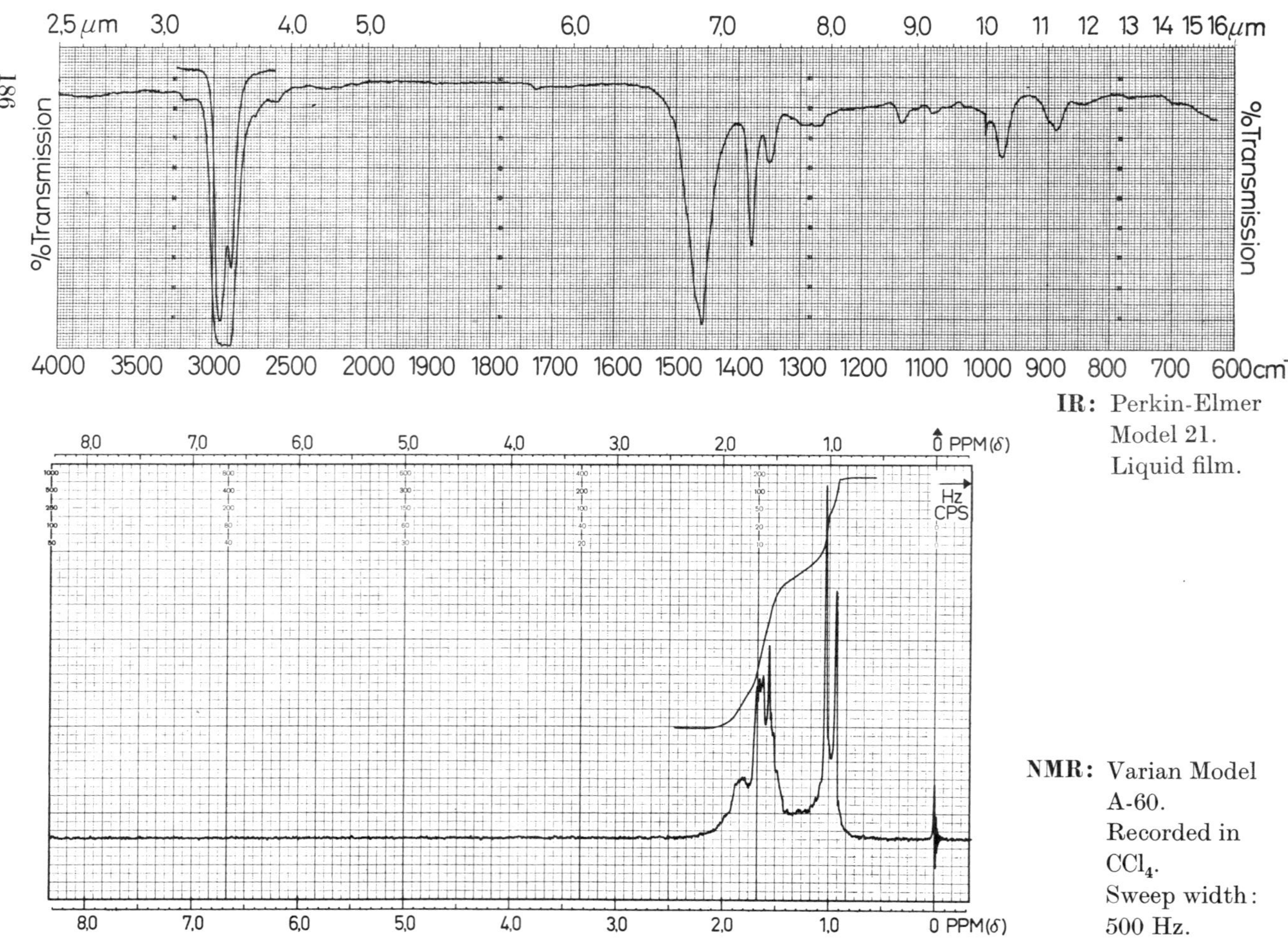

IR: Perkin-Elmer Model 21. Liquid film.

NMR: Varian Model A-60. Recorded in CCl_4. Sweep width: 500 Hz.

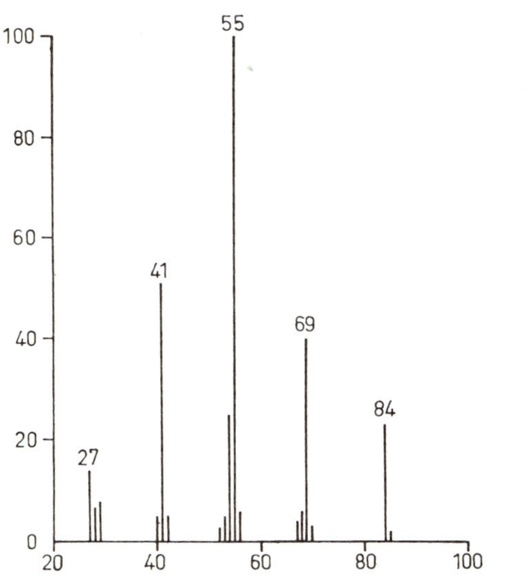

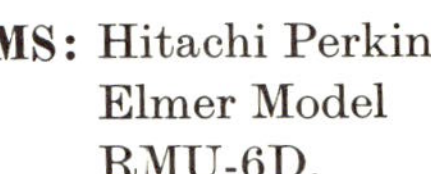

MS: Hitachi Perkin-Elmer Model RMU-6D.

UV: Perkin-Elmer Model 137 UV. Recorded in C_2H_5OH. >210 nm none.

Problem 47

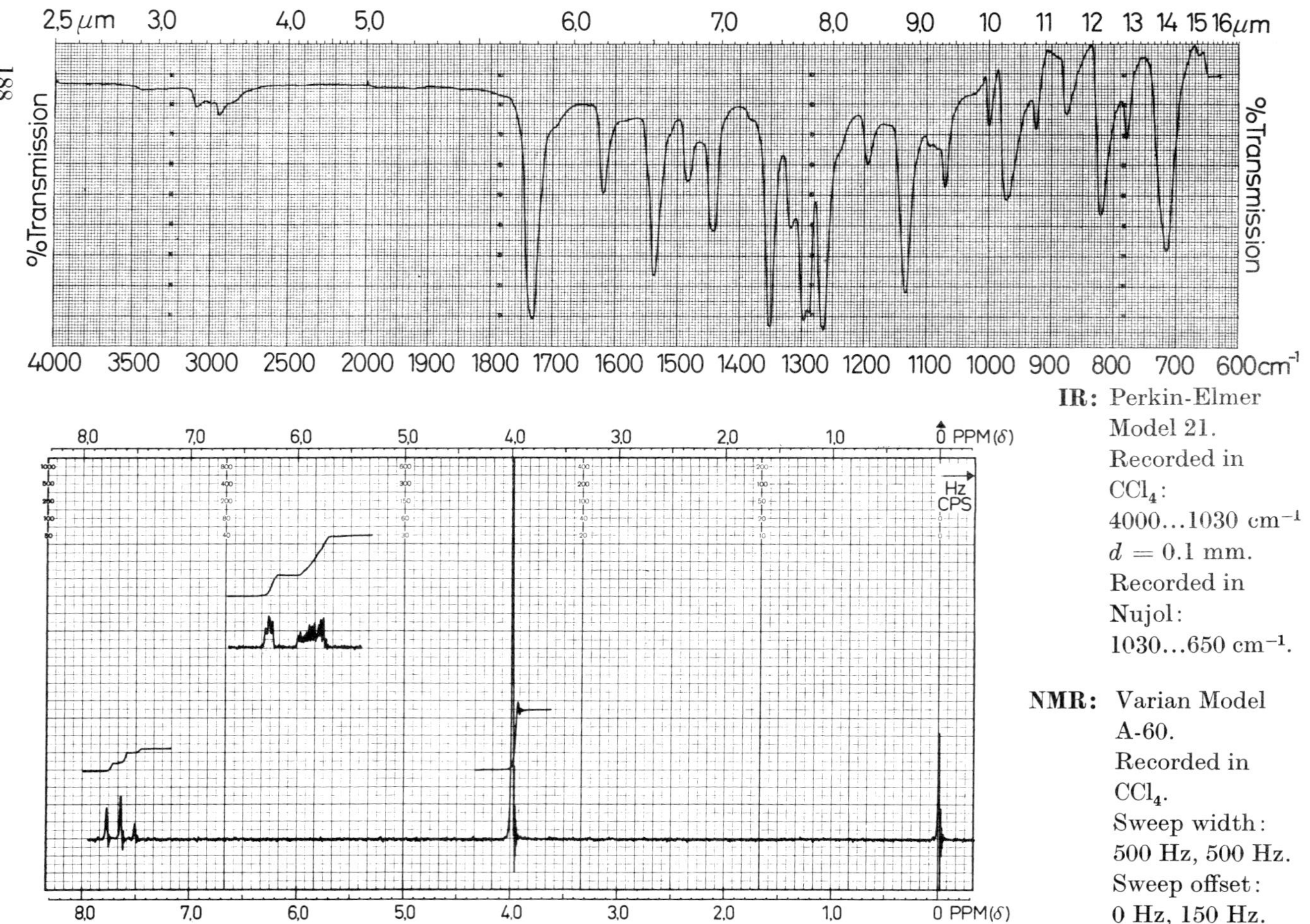

IR: Perkin-Elmer Model 21. Recorded in CCl_4: 4000...1030 cm^{-1} $d = 0.1$ mm. Recorded in Nujol: 1030...650 cm^{-1}.

NMR: Varian Model A-60. Recorded in CCl_4. Sweep width: 500 Hz, 500 Hz. Sweep offset: 0 Hz, 150 Hz.

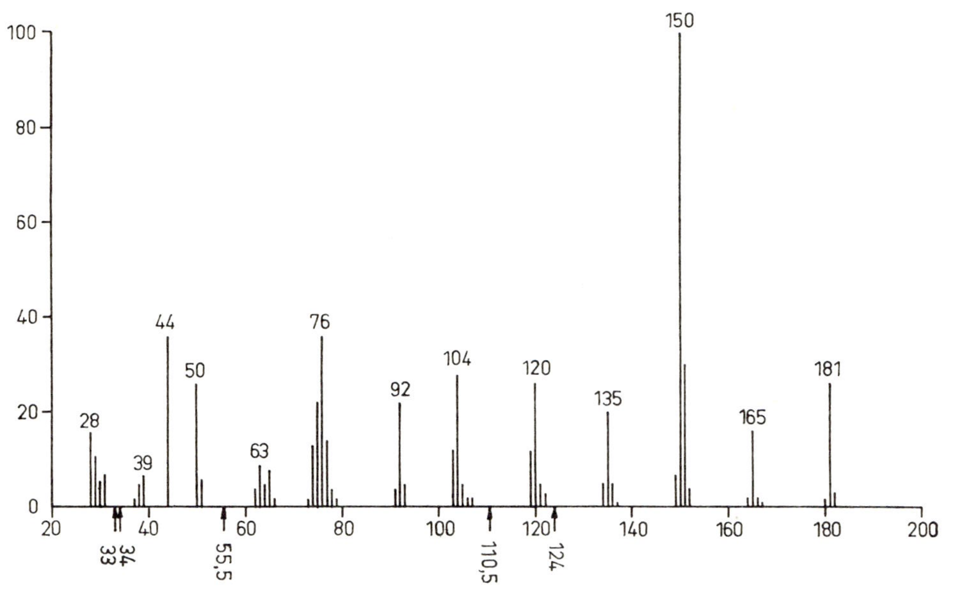

MS: Hitachi Perkin-Elmer Model RMU-6A.

UV: Perkin-Elmer Model 137 UV. Recorded in C_2H_5OH.

λ_{max}	log ε
219 nm	4.4
258 nm	3.8

Problem 48

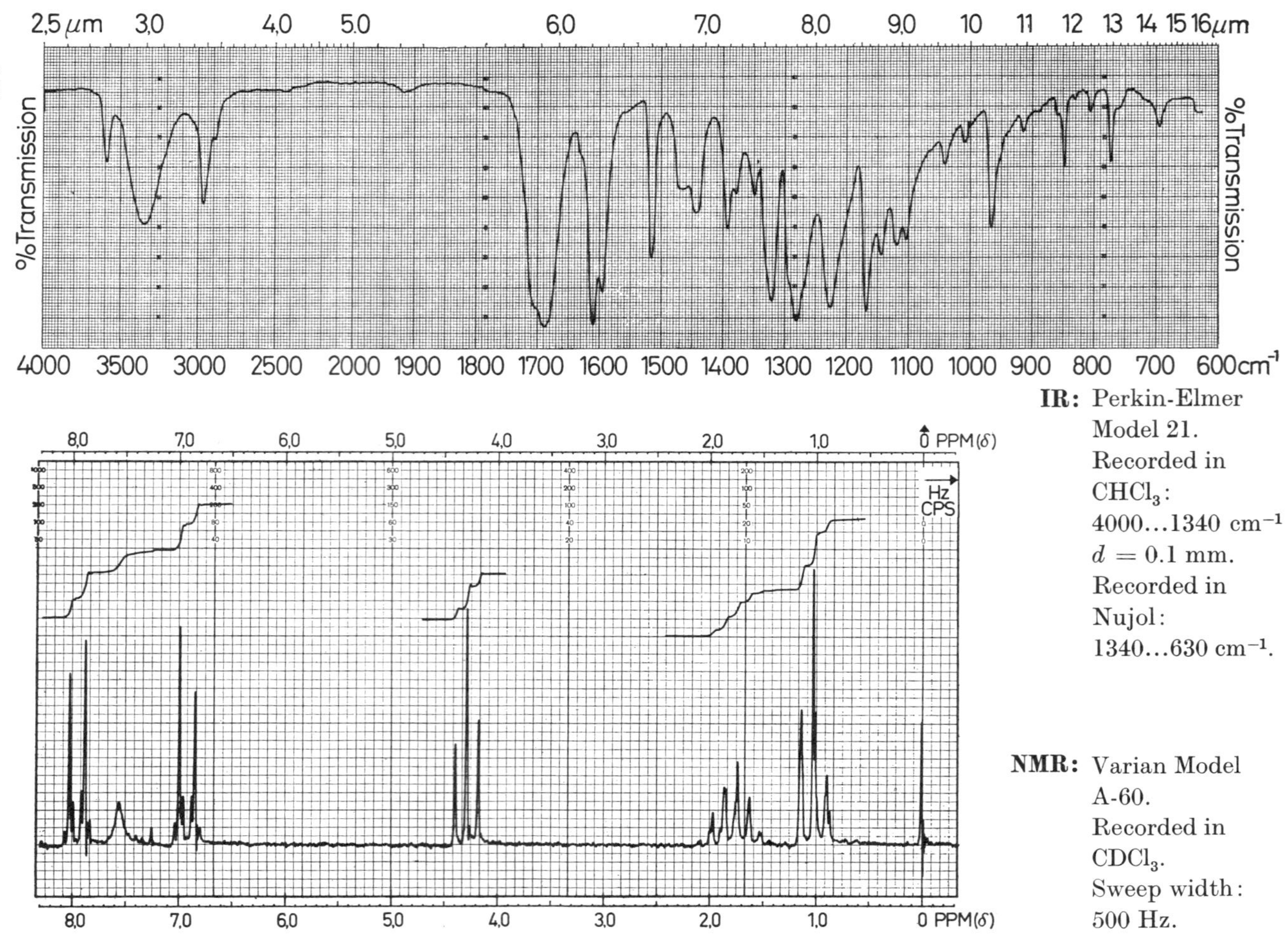

IR: Perkin-Elmer Model 21. Recorded in $CHCl_3$: 4000…1340 cm^{-1} $d = 0.1$ mm. Recorded in Nujol: 1340…630 cm^{-1}.

NMR: Varian Model A-60. Recorded in $CDCl_3$. Sweep width: 500 Hz.

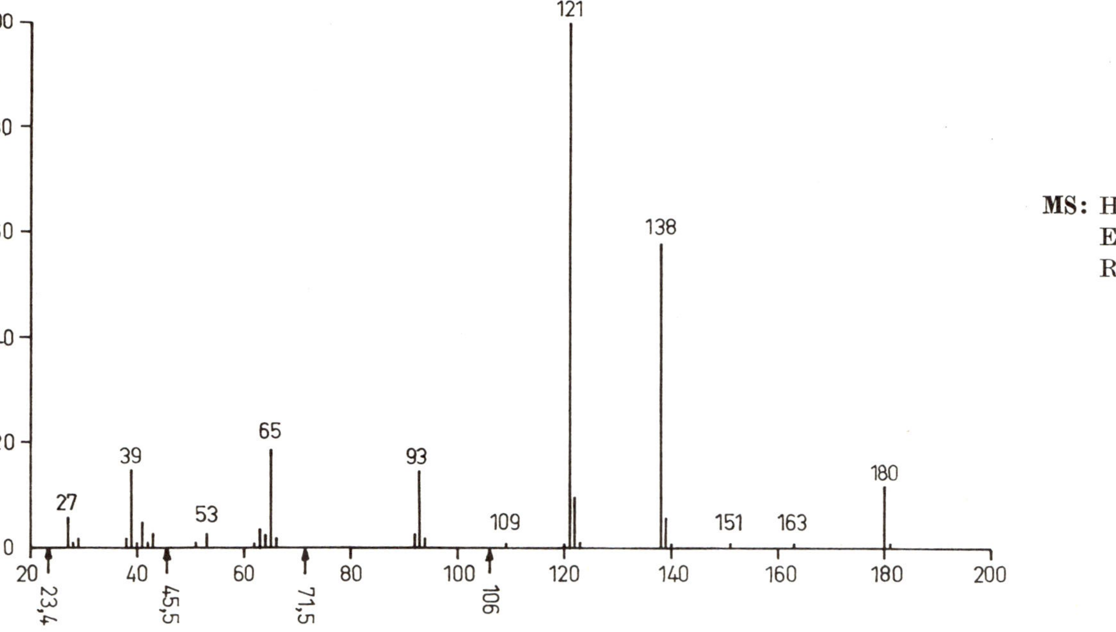

MS: Hitachi Perkin-Elmer Model RMU-6A.

UV: Perkin-Elmer Model 137 UV. Recorded in C_2H_5OH.

λ_{max}	$\log \varepsilon$
258 nm	4.2

Problem 49

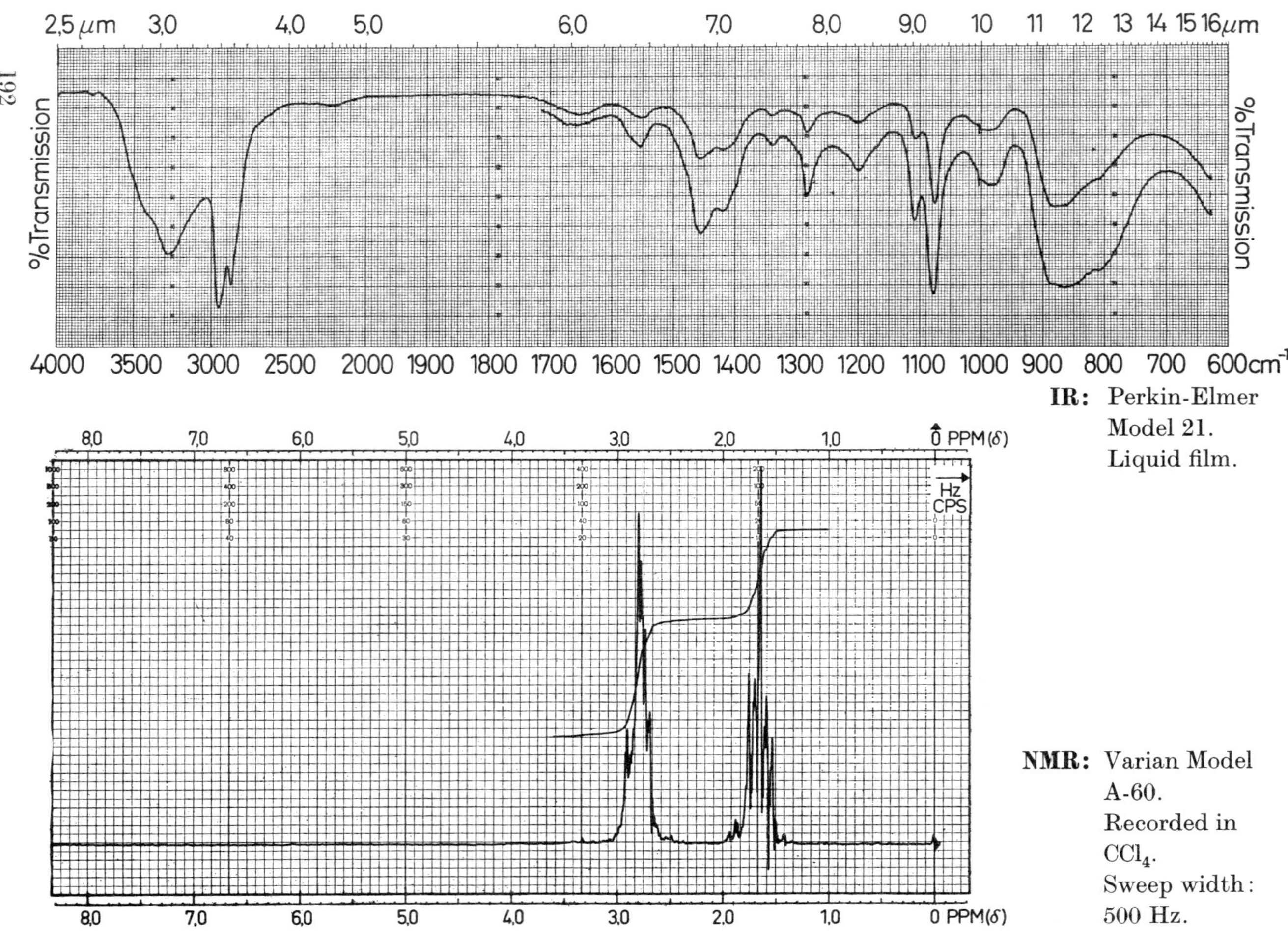

IR: Perkin-Elmer Model 21. Liquid film.

NMR: Varian Model A-60. Recorded in CCl_4. Sweep width: 500 Hz.

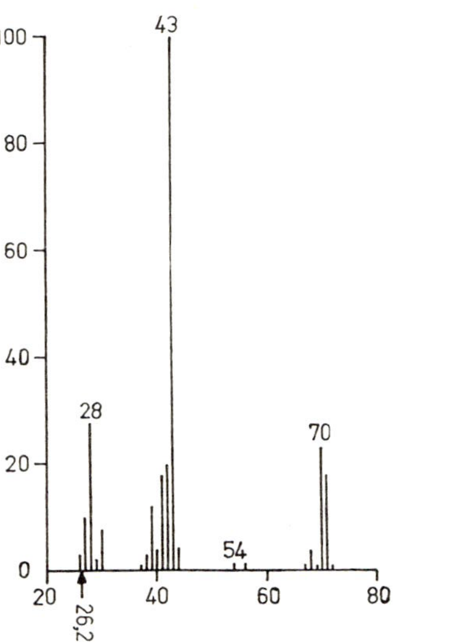

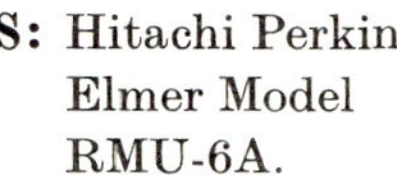

MS: Hitachi Perkin-Elmer Model RMU-6A.

UV: Perkin-Elmer Model 137 UV. Recorded in C_2H_5OH. >210 nm end absorption.

Problem 50

1. 1,1,2,2-Tetrachloroethane
2. 1,3-Dioxolan-2-one. Ethylene glycol carbonate
3. 2-Aminothiazole
4. 2-Ethoxyethanol
5. Methylenesuccinic anhydride, itaconic anhydride
6. Pent-4-enoic acid, allylacetic acid
7. Isopropenyl acetate, acetone enol acetate
8. 4-Methyl-1,3-dioxan
9. (2-Tetrahydrofurfuryl) methanol, tetrahydrofurfuryl alcohol
10. Diphenylamine
11. *p*-Bromonitrobenzene
12. 2,2-Dimethoxypropane, acetone dimethyl ketal
13. 2-Hydroxy-2-methylpropionic acid, α-hydroxyisobutyric acid
14. Ethyl cyanoacetate
15. 3-Aminopyridine
16. Methylmaleic acid, citraconic acid
17. *o*-Benzoylbenzoic acid, benzophenone-2-carboxylic acid
18. 3-Ethoxy-4-hydroxybenzaldehyde
19. Quinoline-2-aldehyde
20. α,4-Dibromoacetophenone, *p*-bromophenacyl bromide
21. Succinimide
22. 9-Benzylfluorene
23. 3-Benzoylpropionic acid
24. 1-Hydroxy-2-naphthaldehyde
25. 2-Ethylbutanal
26. 3,3-Dimethylbutan-2-one, methyl t-butyl ketone
27. 6-Methylheptan-3-one, Ethyl isoamyl ketone
28. 1,8-Cineole. Eucalyptol
29. 2-Ethyl-2-methyl-1,3-dioxolan
30. 1-Bromoheptane
31. 1,1-Diethylpropan-1-ol, triethylcarbinol
32. 2-Methyl-1,4-naphthoquinone
33. 1-Phenylpent-1-en-3-one, ethyl styryl ketone
34. *p*-Nitrophenyl 2,2-dimethylpropionate, *p*-nitrophenyl pivalate
35. Biphenyl
36. *trans*-1,2-Diphenylethylene, stilbene
37. Dibenzyl sulphoxide
38. 2-Phenyl-γ-benzopyrone, flavone
39. 1,3-Diphenylprop-1-en-3-one, benzalacetophenone
40. *N*-Piperoylpiperidine, piperine
41. 4,4′-Bis-(dimethylamino) benzophenone, Michler's ketone
42. Triphenylmethanol, triphenylcarbinol
43. 2,3-Dioxocamphane, camphor quinone
44. 1,2-Diphenylethan-1,2-dione, benzil
45. Pyridine-3-nitrile, nicotinonitrile
46. 3,4-Methylenedioxybenzaldehyde, piperonal
47. Methylcyclopentane
48. Methyl *m*-nitrobenzoate
49. n-Propyl *p*-hydroxybenzoate
50. Pyrrolidine, tetrahydropyrrole

Notes